Retargetable Compilers for Embedded Core Processors

T0135224

Retargetable Compilers for Embedded Core Processors

Methods and Experiences in Industrial Applications

by

Clifford Liem
Laboratoire TIMA &
SGS-Thomson Microelectronics,
Grenoble, France

KLUWER ACADEMIC PUBLISHERS

BOSTON / DORDRECHT / LONDON

A C.I.P. Catalogue record for this book is available from the Library of Congress.

ISBN 978-1-4419-5182-3

Published by Kluwer Academic Publishers,
P.O. Box 17, 3300 AA Dordrecht, The Netherlands.

Sold and distributed in the U.S.A. and Canada
by Kluwer Academic Publishers,
101 Philip Drive, Norwell, MA 02061, U.S.A.

In all other countries, sold and distributed
by Kluwer Academic Publishers,
P.O. Box 322, 3300 AH Dordrecht, The Netherlands.

Printed on acid-free paper

Dedicated to my mother, Mrs. Berty K. Liem,
and my father, Dr. Tik L. Liem.

Thanks for a lifetime of love and encouragement.

Table of Contents

List of Figures

xiv

List of Tables

Foreword

The telecommunications, multimedia and consumer electronics industries are witnessing a rapid evolution towards integrating complete systems on a single chip. System-level integration, combined with extremely short product design cycles, is only possible by implementing large parts of the system functionality in software running on integrated processor cores. Solutions range from general-purpose processor cores available in foundry catalogues, to cost- and power-effective application-specific instruction-set processor cores (ASIPs).

The use of programmable processors is typically divided into two main application classes: computing and embedded systems. Computing applications include desktop computers, notebooks, workstations and server systems. They are characterized by the fact that the end-user can program them, and by the broad range of applications which they perform.

Embedded systems are much more specific in nature and are associated with a dedicated function. Examples are anti-locking brakes, control of electrical appliances and electronic consumer equipment, signal processing for personal audio or wireless terminals, etc. Furthermore, the real-time behavior must conform to very strict requirements. Finally, the correctness of the design is essential due to the potential impact on the surrounding environment or the person using the equipment.

The instruction-set programmable processors used in embedded systems are commonly referred to as embedded processors. These include microcontroller units (MCU), digital signal processors (DSP) and microprocessor units (MPU). Another class of processors encountered in embedded systems is the application-specific instruction-set processor (ASIP). This is a programmable processor that is designed for a specific, well-defined class of applications. It can be seen as a further specialization of the MCU, DSP and MPU classes above. An ASIP is usually characterized by a small, well-defined instruction-set that is tuned to the critical inner loops of the application code.

The focus of this book is nearly exclusively towards tools and methods applicable to embedded processors, an area that has not been given sufficient attention in the past, as most of the work has been targeted to mainstream general-purpose processors in computing-oriented applications.

In this brief introduction, we will address some of the main trends in the use of embedded processors and the applications for which they are designed in order to extract some of the key requirements for embedded software development tools.

Embedded processor trends

While there is a clear trend in processor use for personal computing, with the domination of x86-based architectures and the prevailing use of a single operating system, the consumer electronics, multimedia and telecommunication applications cannot be characterized so easily. Three important factors need to be addressed when considering the future role of embedded processors in these applications:

1. The convergence of computing, communication and consumer electronics. It is likely that the market characteristics of the latter will dominate: extremely short time-to-market, combined with very low costs.

2. The stabilization of the PC market growth.

3. The increasing growth of wireless and multimedia.

This means the applications that will most influence technology evolution in the late nineties and in the early twenty-first century will likely be consumer-oriented, with wireless communication and multimedia the main contributors. These trends have a major consequence on the underlying architectures and the use of embedded processors.

If past history gives any indication of future trends, emerging and future embedded applications like MPEG4, multi-mode wireless, high-definition TV, virtual reality, and interactive 3D games will be expected to be available at competitive prices. This will continue to justify the development of innovative, low-cost processor architectures - and the tools needed to explore these architectures and provide software development platforms.

Embedded systems application trends

It is also important to examine some of the underlying trends of the application requirements associated with embedded systems. The single most important trend is the rapid growth of the complexity of the design:

*   New wireless handsets and base stations will need to support multiple modes, e.g. one or more of GSM, IS-54B digital cellular, the emerging CDMA-based digital cellular, DECT, pager, etc.

*   The integration of communication and fax functions in wireless terminals.

*   Merge of cellular phone and PDA (personal digital assistant) functions.

*   The continued evolution of video coding standards from JPEG, to MPEG1, MPEG2 and eventually, to MPEG4. Each standard evolution is accompanied by significant complexity increase.

*   In multimedia audio, the stereo systems of the eighties are now replaced by complex audio coding standards, e.g. Prologic^{TM}, AC-3, and MPEG2 that support multi-channel 'surround' audio.

As a result, many functions currently in hardware will be performed in software in order to accommodate this increased complexity and evolving standards.

Another important trend is the emergence of standards for embedded applications. In the personal computer area, MSDOS^{TM}, Windows^{TM} and X86 object code have become de facto standards. While the situation is not as simple in embedded applications generally, there are some definite trends in this direction:

1. Multiple sub-system standards are emerging: MPEG2 audio and video, Dolby^{TM} audio (Prologic^{TM}, AC-3, etc.), H.263 videophone, digital wireless (GSM, DECT, IS-54B and CDMA-based standards).

2. These standards are invariably described as ANSI C executable specifications.

3. Furthermore, the chips executing the applications are increasingly being invoked from a standardized applications program interface (API). For example, the new 'DirectX' API designed for PC-based audio and video subsystems, or the common applications programming interface (CAPI) being defined for GSM.

As a result, this frees up the designers in the choice of processor architectures, since they are constrained to the API standard only, and have more freedom on the architecture implementing the functions called via the API. Furthermore, they are not bound by legacy code, and only need to develop a well-defined software function. Once again, this leads to a diversity of processor architectures.

Embedded software needs

The increased complexity of embedded applications, coupled with the predominance of low-cost 8 and 16 bit processors, and the continued importance of ASIPs, leads to some very clear, yet sometimes opposing needs:

1. High-performance compilers for low-cost, irregular architectures, with heterogeneous register structures - such as those found in many MCUs, most DSPs, and nearly all ASIPs.

2. In turn, this implies a compiler development technology that offers:

   • rich data structures, to support the complex instruction sets and algorithmic transformations required to exploit these effectively;

   • extensive search, to explore the numerous register allocation, scheduling and code selection permutations;

   • methods for capturing architecture specific compiler optimizations easily.

3. An environment that supports the rapid development of these compilers, due to the variety of processors to support.

4. The compilers need to be associated with tools like performance profilers, source-level debuggers and in-circuit emulators. These tools need to be retargetable to specific processors with a minimum of effort.

5. For ASIP-based designs, the ability to quickly provide feedback on instruction set selection decisions.

6. Rapid deployment of cycle-accurate instruction-set models.

7. Synthesis of lightweight real-time operating systems (RTOS).

In our opinion, tools that will allow the designer to efficiently transform, refine and map the executable C descriptions onto a cost- and power efficient architecture will provide significant competitive advantage

What this book is about

This book addresses many of the key issues associated with software development tools for embedded systems. It also deals with the critical issue of processor architecture design and analysis.

Chapter 1 discusses embedded cores for today's system-on-a-chip and outlines the evolution of instruction set architectures and the needs for new compilation technologies. This chapter orients the reader to the domain of real-time reactive embedded processors.

Chapter 2 gives an excellent overview of compiler techniques for embedded processors. It includes the basic information needed to understand retargetable compilers. The techniques for compiler construction are reviewed, beginning with traditional compiler techniques for general computing architectures moving to current techniques for embedded processors. The approaches are covered using specific examples from existing compiler systems. Finally, compiler optimizations which are applicable to embedded processors are reviewed.

Chapter 3 deals with industrial developments of retargetable compilers. This chapter outlines two examples of retargetable compiler systems. The first is Code-Syn, a model-based compiler developed at Bell-Northern Research / Nortel. The second is the FlexCC rule-driven compiler used at SGS-Thomson Microelectronics. The functionality of these compilers is described, with an emphasis on the unique features of each approach. The chapter concludes with an assessment of the two approaches.

Chapter 4 provides valuable feedback of practical issues from an industrial viewpoint. This chapter shows the large industrial experience acquired by the author. It presents the fundamentals of pragmatic approaches to retargetable compilation. This begins with the treatment of engineering issues when developing a firmware development environment for embedded processor systems such as language support and source code style writing. Also included are other issues such as strategies for compiler validation and software debugging.

Chapter 5 is devoted to new methods and tools for address calculation. These techniques are critical for the design of embedded digital signal processing (DSP) systems. This chapter describes retargetable transformations which map high-level constructs onto parallel executing post-modify address calculation units. Discussion of an extendable model for address calculation units (ACUs) is presented with supporting algorithms.

Chapter 6 analyzes the industrial application of the new advanced compilation methods for several specific processors designed for recent products in multimedia and telecommunications. This is an outline of experiences building compiler environments with design teams at Bell-Northern Research / Nortel, SGS-Thomson Microelectronics, and Thomson Consumer Electronic Components. A description of the architectures and unique challenges are presented, followed by an outline and evaluation of the engineering effort used to develop the embedded software development environments. The examples cited are for recent products in multimedia and telecommunications.

Chapter 7 investigates instruction-set design and exploration through the combination of retargetable compilers and new utilities for architecture analysis. These tools permit a designer to analyze the relationship between application code and the instruction-set of the processor.

Why this book is of interest

The treatment is both from a theoretical viewpoint, where an up-to-date literature review is presented; and from a practical viewpoint, where the application of retargetable software development tools is described in the context of industrial embedded systems. It is the dual nature of this book that makes it of high value to the researcher and tool developer alike.

Finally, an important characteristic of this book is that many of the tools and methods described have been applied successfully to leading industrial applications in DSP, multimedia and low-cost embedded control. Real products currently in volume production have benefitted from these contributions. Of course, some theoretically interesting approaches needed to be traded off against more pragmatic solutions. It is precisely these trade-offs which are interesting to the engineer or manager in the context of the introduction of new design technology in industrial design flows. In any case, a substantial productivity increase has been demonstrated which led to faster time to market - the yardstick against which most new approaches are measured.

Pierre G. Paulin & Ahmed A. Jerraya

Preface

Embedded core processors are becoming a vital part of today's system-on-a-chip in the growing areas of telecommunications, multimedia and consumer electronics. This is mainly in response to a need to track evolving standards with the flexibility of embedded software. Consequently, maintaining the high product performance and low product cost requires a careful design of the processor tuned to the application domain.

With the increased presence of instruction-set processors, retargetable software compilation techniques are critical, not only for improving engineering productivity, but to allow designers to explore the architectural possibilities for the application domain.

This book overviews the techniques of modern retargetable compilers and shows the application of practical techniques to embedded instruction-set processors. The methods are highlighted with processor examples from industry used in products for multimedia and telecommunications. An emphasis is given to the methodology and experience gained in applying two different retargetable compiler approaches in industrial settings. The book also discusses many pragmatic areas such as language support, source code abstraction levels, validation strategies, and source-level debugging. In addition, new compiler techniques are described which support address generation for DSP architecture trends. The contribution is an address calculation transformation based on an architectural model.

As a natural complement to the compiler techniques which have been presented, new utilities for the design of embedded processors are described. As the lifetime of an embedded processor is often rich with architectural variations and hardware reuse, these aids provide ways of analyzing the match between application code and the instruction-set. Two tools allow the designer to obtain both static and dynamic feedback on the fit of the processor in the application domain. These tools allow the designer to explore the architecture space as well as the algorithm execution.

The intended audience of the book includes embedded system designers and programmers, developers of electronic design automation (EDA) tools for embedded systems, and researchers in hardware/software co-design.

Acknowledgments

A great number of people have contributed to the work contained in this book and I would like to thank each one of them for their help and support. To Pierre Paulin, thank-you for nearly six years of productive and interesting cooperation. In addition to being a role model for my technical career, you've also been a supporting friend, always looking out for my best interests. To Ahmed Jerraya, thanks for the open welcome into your team. You've motivated me in more ways than I could describe.

To Marco Cornero, thank-you for your diligent corrections to this text, not to mention your technical contributions to the projects. You've been a pleasure to work with both professionally and personally. To François Naçabal, thanks for your support through my time in France both as a friend and colleague. To the rest of the Embedded Systems Team at SGS-Thomson: Miguel Santana, Etienne Lantreibecq, Frank Ghenaissa, Michel Favre, Frédéric Rocheteau, and Thierry Lepley, thank-you for many positive discussions both in groups and one-on-one.

To the TIMA Laboratory and particularly the Systems Level Synthesis group, thank-you for making my stay enjoyable and productive. I especially thank (in no particular order) Maher Rahmouni, Imed Moussa, Gilberto Marchioro, Jean-Marc Daveau, Carlos Valderrama, Jean Fréhel, Julia Dushina, Wander Cesario, Richard Pistorius, Hong Ding, Tarek Ben-Ismail, Mohamed Romdhani, Mohamed Abid, M. Ben Mohammed, Polen Kission, and Ricardo de Oliveira Duarte for many technical and personal discussions. Furthermore, I thank Phillippe Guillaume for his careful corrections of the French summary of this writing. And to the rest of the TIMA/CMP Laboratory headed faithfully by Bernard Courtois, I thank you for a comfortable working environment with such a qualified staff. I give particular thanks to Isabelle Essalhiene-Schauner, Corinne Durand-Viel, Chantal Benis and Patricia Chassat.

To the IVT design team of SGS-Thomson in particular Michel Harrand, Olivier Deygas, José Sanches, and Elisabeth Berrebi, thanks for the many fulfilling cooperations. To the MMDSP design team at Thomson Consumer Electronic Components, I thank Laurent Bergher, Jean-Marc Gentit, Xavier Figari, and Jean Lopez for giving me the opportunity to know and work with you. Finally, to the DPG Video design team of SGS-Thomson: Richard Chesson, Giovanni Martinelli, and Min Xue, thanks for the openness in using my prototype design tools.

I would like to thank Central R&D of SGS-Thomson Microelectronics including Jo Borel, Yves Duflos and Jean-Pierre Moreau for the support of this project. I would also like to thank Bell-Northern Research/Nortel including Dave Agnew and Terry Thomas for their support of this project. I especially thank Donna Gervais for her tireless help throughout this co-operation. I would also like to thank my former colleagues at Nortel for their technical contributions which sparked the beginnings of much of this work: Trevor May, Shailesh Sutarwala, Francis Langlois, and Chris Donawa.

xxvi

And to my wife Christine Sauvé-Liem, your love and support was the final link that gave me the motivation to complete this work. Thanks for always being by my side. I also give you a special thanks for help in the design of the book cover. You have a wonderfully creative gift.

Finally, I thank You, my Lord and Saviour, Jesus Christ, for your unending love and grace all the days of my life.

Chapter 1:   Introduction

1.1 Embedded processors for today's system-on-a-chip

As microelectronic fabrication capabilities evolve to astounding levels of submicron technology, more and more functions can be integrated on-chip. Industries such as telecommunications and consumer electronics are witnessing a rapid evolution to entire systems being placed on a single die.

At the same time, demands upon system designs are mounting. The standards organizations which set the quality level whereby microelectronic systems are judged are continually in flux. This constant evolution in standards causes churn in hardware designs and can sometimes mean the decision for costly redesigns.

In effect, programmable processors are becoming increasingly present as a design solution. An example of this trend is the SGS-Thomson Integrated Video Telephone. While the Video CODEC version of the chip (the STi1100 [92]) contains two programmable cores, the current version being designed contains five embedded cores (see Section 6.2.). The most attractive reason for the use of a programmable solution is the ability to track the evolving standards using software for late design changes. Furthermore, with a custom instruction-set core, high speed, low cost, and low power are not compromised. Figure 1.1 depicts some of the decisions which make programmable processors a compelling design style.

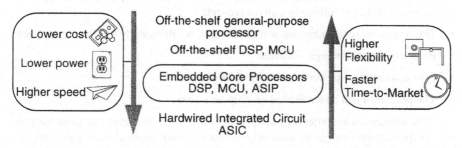

Figure 1.1 Embedded core processors: solutions to conflicting requirements.

Additionally, higher levels of integration are encouraging design practices such as the reuse of macro blocks. Whether these blocks are hard (i.e. netlists and/or layouts) or soft (i.e. synthesizable VHDL, Verilog), embedded processors are a conve-

1

nient manner to reuse intellectual property. The trend is especially clear within companies where embedded processor designs are reused and evolved beyond their original intention.

Finally, an enticing feature of an embedded processor is the ability to carry out concurrent engineering practices between hardware and software design teams. The instruction-set of the processor serves as the rigid contract between the two teams to carry a product to market in an efficient time cycle.

While it can be established that the use of embedded processor cores is advantageous, the design flow which supports their use is much different from the standard hardware design flow and even the design flow for general purpose processors. A key technology in the design for embedded processors is retargetable compilation; however, the techniques in this area are just beginning to appear.

We define an embedded processor as the principal component of an embedded system. An embedded system can be defined by describing it's main properties (Camposano and Wilberg [8]):

- The system performs a dedicated function.

- The system's real-time behavior must conform to very strict requirements.

- The correctness of the design is essential due to the impact on the surrounding environment.

The focus of the entire manuscript is on the design tool needs for deeply embedded processors used in embedded systems. Examples include DSPs (Digital Signal Processors) and MCUs (microcontrol units) used in the consumer electronic domains of multimedia. and communications. A processor in these example systems typically has many or all of the following criteria present:

- The processor runs embedded software which is infrequently modified (i.e. *firm*).

- Products are sold in high volumes.

- Low cost and low power are of critical importance.

- The processor can be embedded on-chip as a core.

The discussion hereafter will neither include the requirements and tools for general-purpose microprocessors used in workstations and personal computers, nor the requirements for highly parallel computers. Both of these types of machines have a very different set of design tool constraints and conditions when compared to embedded core processors.

This chapter begins with a review and look at the trends of the instruction-set architectures used in today's embedded systems in the application domains of telecommunications and multimedia. Following the discussion of the processor architectures, an overall picture of the needed design tools for embedded processors is discussed. The chapter then concludes with a summary of the objectives and organization of the rest of the book.

1.2 Embedded instruction-set architectures: What's new?

1.2.1 The evolution of embedded processor architectures

The evolution of embedded processors begins with a complex set of design principles arising from the general computing area. As these principles are brought into the world of real-time reactive systems, the constraints of the application areas affect the characteristics of the architectures.

VonNeumann and Harvard Architectures. One of the simplest designs of an instruction-set processor is the VonNeumann architecture as shown in Figure 1.2. This design is characterized by a single memory containing both the program instructions and the data to be processed. The controller for this type of architecture is straight-forward and synchronism between instructions and data is simple. However, when considering speed the data and address calculations which are done on the ALU are hindered by a significant bottleneck on the shared instruction and data bus. Furthermore, the register file shares this same bus. Another architectural point is that data and instructions are forced to have the same bit-width.

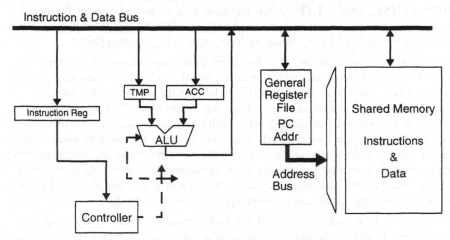

Figure 1.2 Simplified VonNeumann instruction-set architecture

To relieve some of the weaknesses of the VonNeumann architecture, the Harvard architecture was introduced. An example is shown in Figure 1.3. The major difference between the two design styles is the separation of program and data memory. This allows the use of two separate busses improving the overall speed of the unit. Many variations can be added to the basic Harvard architecture including such examples as separate data and address register files, multiple data memories, and the addition of functional units like an address calculation unit (ACU).

Data Bus

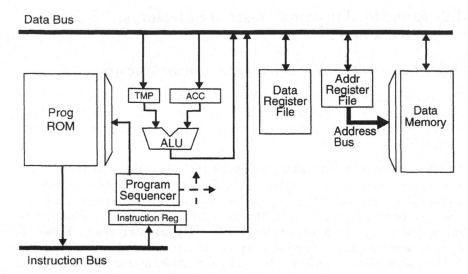

Instruction Bus

Figure 1.3 Simplified Harvard instruction-set architecture

RISC, CISC, and VLIW. Although there is a wide variation in the details, microprocessors have evolved from two broad principles: *CISC* (Complex Instruction-Set Computer) and *RISC* (Reduced Instruction-Set Computer) [76].

The general principles of CISC dictate the direct implementation in the machine of large, complex operations which could be found in a high-level language. Large pieces of functionality are made directly available in the hardware. An instruction in a CISC machine may contain many addressing modes available for the same ALU operation. For example, an indirect offset addressing mode may be available which can directly implement a memory reference from an array. The machine would calculate the correct address at run-time to retrieve the data. In general, the instruction-sets are rather large since many combinations of instructions are present. An instruction is usually based upon a *data-stationary* concept which may take any number of cycles to execute until the calculation is complete (Goossens et. al. [36]).

In principle, the advantages of a CISC architecture is a simplified compiler mapping from a high-level language to instructions since many complex operations are directly available. Consequently, the CISC can also achieve a relatively high program code density. However, the simplicity of the compiler mapping is debatable, since the exploitation of complex operations is not always straightforward. Another disadvantage of the CISC mechanisms is a rather high amount of hardware complexity including the decoding and implementation of a large number of addressing modes.

In contrast, the RISC principles adopt a reduction of the instruction-set size to the bare minimum allowing a simplification of the hardware implementation and control. For example, in RISC machines, loads from memory and stores to memory are separated from ALU operations by intermediate registers. This implies the use of very simple addressing modes which allow single-cycle execution of all the operations. This memory hierarchy also improves data throughput by allowing operations to be executed in a *pipeline*, where a second operation can be started before the first is actually complete. The idea can also be extended to other functional units such as an ACU (Address Calculation Unit) which independently executes addressing operations in a pipelined fashion. Instructions in a RISC machine are usually based upon a *time-stationary* concept, whereby all instructions take the same time to execute (Goossens et. al. [36]).

The advantage of a RISC machine is a much simpler hardware control implementation and a smaller instruction-set. It also allows faster execution with the possibility of pipelining. However, compilers must be able to manage smaller individual operations. For example, an array reference must be separated into individual address calculations in contrast to a CISC principle where the addressing mode already exists. On the other hand, in practice, the RISC principle has actually simplified the compiler design since a smaller instruction-set is easier to manage than a large instruction-set.

The *VLIW* (Very Long Instruction Word) concept is a principle which extends RISC fundamentals permitting a maximum of parallelism in a single machine cycle. The instruction-word is encoded in a way that allows operations to be executed in a manner independent from one another. This notion is known as *orthogonality*. Instruction widths can vary from a fairly narrow 61 bits for embedded applications (the TCEC MMDSP in Section 6.3) to 759 bit words for highly parallel machines! An example is an IBM VLIW architecture [120] shown in the diagram of Figure 1.4. The control logic for a VLIW is relatively simple compared to other high performance processors since there is no dynamic scheduling or reordering of operations as in many *superscalar* processors.

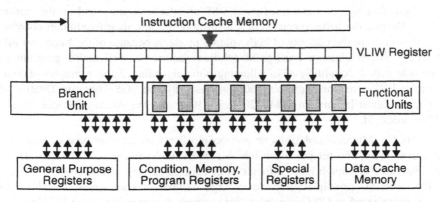

Figure 1.4 Example VLIW architecture with multiple functional units which execute in parallel [120]

For embedded real-time processors, a disadvantage of a VLIW is the high use of program memory. A VLIW program can have many poorly used bits stemming from the orthogonality in its wide instruction-word. This corresponds to unneeded memory size overhead which is an expensive consideration for a system-on-a-chip.

Digital Signal Processors and Microcontrollers. *DSPs* (Digital Signal Processors) are a type of architecture specialized for data intensive applications. They are characterized by certain functional blocks which allow the processors to function efficiently on typical signal-processing algorithms. Some examples are the algorithms for digital filters (FIR (Finite Impulse Response), IIR (Infinite Impulse Response), fast fourier transforms (FFT), noise elimination, and echo cancellation. The characteristic functional blocks of these architectures include multiply-accumulators (MACs), address calculation units (ACUs) with modulo and bit-reverse addressing modes, barrel shifters, and multiple memories.

The majority of today's commercial DSPs are based upon Harvard and RISC properties to meet the most stringent constraint of performance. Parallelism is also a principal performance gaining factor which naturally leads to VLIW architecture considerations. However, DSPs cannot afford costly wide instruction words, which implies the use of highly encoded instructions. An example of DSP encoding restrictions is illustrated in Section 1.3.2. These machines still allow the full parallel execution of specifically chosen instructions important for signal processing. At the same time, orthogonality of the instruction-set is greatly diminished. However, the program memory savings is an important gain for the hardware.

DSP architectures are also characterized by heterogeneous and distributed register structures. Registers are first associated directly with the input and output of particular functional units and secondly reserved for special purposes. This is again a performance gain when compared to architectures with large general register files. Often, the capability of *coupling* registers for large data-types is also present, especially for functions like multiply-addition to preserve the precision of the calculations (see Figure 1.7).

Both floating-point and fixed-point DSP architectures are found on the market today. Despite the better precision of a floating-point unit, their fixed-point counterparts are used in the majority of high volume products because of the huge cost difference. Nevertheless, a fixed-point solution requires a well coded program to compete with the quality level of a floating-point algorithm. Some examples of commercial DSPs are: the Motorola 56000 series [78], the SGS-Thomson D950 core [91], the Texas Instruments TMS320 series[100], and the Analog Devices ADSP-21xx series [5].

MCUs (Microcontrol Units or microcontrollers) are another type of real-time reactive instruction-set processor which are oriented toward control tasks. These architectures are much more difficult to categorize as they can be based on either CISC or RISC principles; however a large number of commercial 8 and 16 bit devices are based on CISC principles. For example, it is common for an MCU to contain over 10 addressing modes and over 100 instruction types. What sets them apart from microprocessors is the general low cost and the explicit functions for memory and input/output control tasks used for real-time interaction in an environment. Nevertheless, microprocessors have also been used in real-time environments, yet they generally have significantly higher costs. Some examples of commercial microcontrol units are the Motorola 68HC05, 68HC11, MPC500 [117], the SGS-Thomson ST6, ST7, ST8, ST9 [95], and the Texas Instruments TMS370 family [121].

1.2.2 Embedded processor architectural directions

Compounding an enormous variety of instruction-set architectures being used today, one clear trend in embedded processors is the support of architectural variations. Looking through the portfolio of a major semiconductor vendor's offerings is more overwhelming than looking through a clothing catalogue with different colors and sizes! For example, the Motorola 68HC11 MCU is categorized by a first set labelled A through P, then further subcategorized resulting in over 50 members. The SGS-Thomson ST9 MCU is offered in a multitude of packages (Dual in Line Plastic Package, Plastic Leaded Chip Carrier Package, Window Ceramic Leaded Chip Carrier, Plastic Quad Flat Pack, etc.), with a variations on pin input/output (32, 36, 38, 40, 56, 72), and a large variation on memory configuration (ROM 8/16K, EPROM 16/32K, RAM256/1280K, EEPROM). Of course, there is also the option of different peripherals of various shapes and sizes: Multifunction Timers, DMA (Direct Memory Access), Analog-to-Digital Converters, etc.

This trend continues for DSPs. For the Motorola 56K series, they are categorized in five main families: the DSP56000 for digital audio applications, the DSP56100 for wireless and wireline communications, the DSP56300 for wireless infrastructure and high MIPs applications including Dolby AC-3 encoders, the DSP56600 for wireless subscriber markets, and the DSP56800 for low cost consumer applications [117]. Another example, the SGS-Thomson D950 core has several memory configurations as well as a configurable coprocessor interface [91]. A set of instructions are set aside to communicate with a coprocessor which may be added to customize the hardware to a particular application algorithm.

It is clear that for the embedded processor market, it is not enough to have a product with a fixed architecture. A solution is competitive because it is specialized for the application domain and the architecture is refined for the type of algorithms to be executed.

The concept of architecture customizing has been taken even further in some companies such as Philips which have designed the flexible EPICs DSP core [111] for a range of products including digital compact cassette players (DCC), compact disc players, GSM mobile car telephones, and DECT cordless telephones. Flexibility in the EPICs architecture includes the customization of word-lengths, peripherals, memory types, memory dimensions, and register sets.

In high volume products, it is apparent that the concept of architecture customizing is a principal competitive factor. By consequence, the architecture trend is the move toward dedicated processors built using flexible variations on a theme. This type of architecture is known as an *ASIP* (Application Specific Instruction-Set Processor).

1.3   Tools for embedded processors: What's needed?

1.3.1 Ask the users

For the large majority of real-time embedded firmware, assembly is the common source language [86]. While there is a general awareness that high-level languages bring many more benefits including readability, portability, and easier maintenance, the current state of compiler technology for embedded processors is less than acceptable. For example, the DSPStone benchmarking activities [112][113] have demonstrated the low efficiency of commercially available DSP compilers including the Motorola 56001, the Analog Devices ADSP2101, the AT&T DSP1610, the Texas Instruments TMS320C51, and the NEC uPD77016. All but the last two compilers are retargets of the GNU gcc compiler (described in Section 2.1.2). For these processors, compiled code for a set of DSP algorithms was shown to run from 2.5 to 12 times slower than hand-coded algorithms! For a designer of real-time systems, a 20% performance overhead is typically the tolerance limit [86]. This makes these commercial compilers unusable.

As embedded systems become more sophisticated, the amount of legacy code in assembly programs becomes so large that code management becomes a serious issue. Consequently, embedded system programmers do recognize a real need for compiler technology. This was demonstrated in a survey of designers for telecommunication systems [83][86], indicating that the greatest need for embedded processors is the presence of efficient software compilers.

While the compiler technology for embedded processors has not yet advanced to an adequate level, an embedded system does provides some unique opportunities for the compiler developer. Unlike for general-computing systems, a program for an embedded system is well simulated and validated on a host platform before being downloaded to the final embedded system (Figure 1.5). This offers the opportunity to

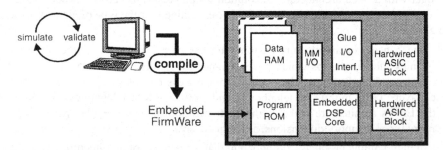

Figure 1.5 Host simulation and validation before final download to the embedded system

make use of such items as a host compiler, profiling tools, and execution based optimization strategies. Of course, the time needed for thorough simulation is always an open problem. Nevertheless, simulation remains an integral part of any embedded software development cycle.

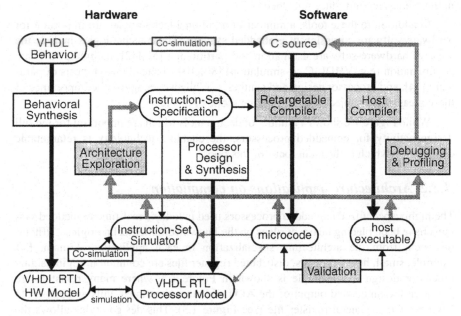

Figure 1.6 Design tools for embedded processors.

Figure 1.6 shows the full picture of the design tools that we envision to be needed for systems containing embedded instruction-set processors. The heart of the hardware-software design flow is a retargetable compiler which is reconfigured by means of an instruction-set specification. Modifications to this specification serve as a method to explore the effect of making architectural changes on the performance of the C source algorithm. Furthermore, the specification could also be used to generate an instruction-set simulation model and a hardware description of the processor itself.

For this design flow, the key technologies are the compiler techniques which map C algorithms to microcode by means of an instruction-set specification. A review of compiler techniques is presented in Chapter 2, with contributing techniques and methodologies in Chapters 3, 4, and 5.

The use of a host compiler (e.g. workstation or personal computer) serves multiple purposes. The first is early functional verification of the source algorithm even before a processor design is available. The second purpose is validation of the targeted compiler. These subjects are discussed in detail in Chapter 4. By consequence, the presence of both a retargetable compiler and host compiler also allows further possibilities. They can be used for debugging in various forms (Section 4.4) and for architecture and algorithm exploration. For example, the knowledge of which instructions are used by a source algorithm in both a static and dynamic fashion is useful for the refinement of the system performance. Both processor hardware and algorithm software may be refined to the algorithm needs. Tools which aid a designer in these areas are introduced in Chapter 7.

In addition to these tools, a number of additional technologies are important for hardware-software co-design of embedded systems. These technologies include the areas of hardware-software estimation and partitioning [51][42], hardware-software co-simulation (e.g. VHDL-C co-simulation) [80][104], behavioral synthesis of hardware [14], and processor design and synthesis. While we recognize the importance of these areas, these subjects are beyond the scope of this text.

While Figure 1.6 shows a number of design tools which are important for the full design activity for embedded processors, the enabling technology is retargetable compilation which is the main focus of this book.

1.3.2 Architecture implications on compilation

The highly specialized embedded processors used in today's real-time embedded systems have been placing heavy burdens on the known compiler technologies. Difficulties stem from the architecture specialization in each application domain. For example, small, heterogeneous, distributed register files are common in digital signal processor design. An example is shown in Figure 1.7, where many registers are placed at the inputs and output of the ALU and other functional units as opposed to having a large, general register file (see Figure 1.3). This design styles allows the instructions to be encoded in a manner which keeps the instruction width to a minimum. However, it also means that registers are used for specific roles and sometimes overlapping roles. For example, in Figure 1.7 S1 could be a special register used solely for bit-shifting operations. Another example would be the *coupling* of the output registers A1 and A2 to be used for a double-precision data-type, as well as the use of each register separately for single-precision data-types.

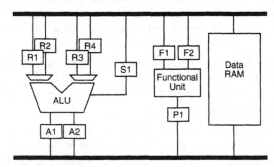

Figure 1.7 Example of heterogeneous, distributed register files

Specialized registers imply that a compiler needs to treat registers based on their function in a certain context. In contrast, the trend in general-purpose processors is to provide a large number of registers which can be used for any function. This is done primarily to simplify the compiler task. Now, as embedded applications change the requirements of a programmable architecture, compilation technologies are called to

Multiply
$$P = Left\_src * Right\_src (Rnd)$$

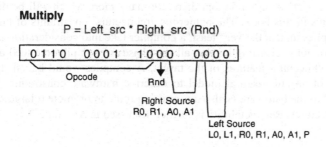

Multiply-Accumulate with 2 indirect Register Loads
$$A \mathrel{+}\mathrel{-}= P, P = Left * Right (Rnd), Lx = *AX + IX, Ry = *AY + IY$$

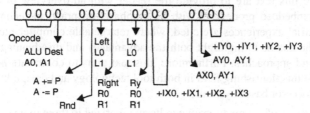

Figure 1.8 Instruction encoding for the SGS-Thomson D950 DSP core.

keep pace.

Another sizable challenge for DSP compilation is dealing with architecture restrictions as a result of instruction-word encoding. An example of tight encoding restrictions is shown in Figure 1.8 for the 16-bit SGS-Thomson D950 DSP core. Two instruction types are shown: the simple `Multiply` instruction and the `Multiply-Accumulate with 2 indirect Register Loads`. Notice the differences which distinguish the format of each instruction. The opcodes are of different widths: 10 and 3 respectively. However, the greatest impact is the difference in allowable register usage for each instruction. In the `Multiply` instruction, the right source may be any of the 4 registers R0, R1, A0, or A1, and the left source may be any of the 7 registers: L0, L1, R0, R1, A0, A1, or P. However, for the `Multiply-Accumulate` instruction, the left source for the multiply operation may only be one of two registers L0 or L1, while the right source may only be: R0 or R1. Looking closely at the instruction word, it is clear that the designer has a heavy encoding constraint if he wants to offer a large amount of parallelism.

The implications of these forms of encoding restrictions on the compilation techniques are enormous. For a compiler to make use of the `multiply-accumulate` instruction in the previous example means that it must be treated by all the basic phases of compilation including: instruction-set matching/selection, register allocation/assignment, and scheduling/compaction. Furthermore, optimizations such as loop pipelining are of extreme importance. These and other existing compiler techniques are reviewed in Chapter 2.

While we have just touched on some of the architectural considerations of embedded processors on compilation, there are many more which will be discussed at various points in this book. The numerous architectural constraints of embedded processors implies that, at the very least, a compiler take into consideration all of the hardware restrictions. This implies that all the phases of compilation need a knowledge of the architectural features of the target. A compiler would benefit from the incorporation of an architectural model to describe hardware constraints. Furthermore, a model of the hardware is also a promising route to promote retargetability to varying architectures. Retargetability is further discussed in Section 2.2.

1.4 Objectives, contributions, and organization

The objectives of this text are to provide the reader with an overview of compiler technology for embedded processors with an emphasis on practical techniques. A number of industrial experiences are cited, where retargetable compiler methodologies are used. The goal is to highlight both the advantages and disadvantages of the methodologies and approaches. Furthermore, this text aims to contribute new ideas and techniques to this flourishing field in both tool technology and design know-how for embedded processor based systems.

The contributions of the manuscript can be summarized in three main categories:

- experiences and methodologies in compiler approaches for embedded processors in the context of industrial products for telecommunications and multimedia.
- a new compilation approach to address generation for Digital Signal Processors based on an architectural model.
- a set of tools which allows the designer to explore the fit of a set of applications on a processor in light of an architecture evolution or reuse.

The organization of the rest of this book is as follows: Chapters 2-6 describe compiler methodologies for embedded processors and their application in industrial case studies. This begins in Chapter 2 with an overview of traditional and embedded processor compiler techniques. Chapter 3 describes two retargetable compiler systems developed in industry for embedded processors. Chapter 4 discusses a number of practical issues which are needed in any methodology incorporating a compiler for an embedded processor. This is followed in Chapter 5 by a new approach to a compiler transformation specifically for address generation: a critical part of compilation technology for DSPs. Chapter 6 describes a number of case studies with industrial processors, using the techniques presented in the previous chapters.

Chapter 7 presents architecture and algorithm exploration tools which are complementary to an embedded processor development environment. Finally, Chapter 8 presents a wrap-up of the contributions of the book followed by a reflective outlook on the horizon.

Chapter 2: An Overview of Compiler Techniques for Embedded Processors

The challenge of constructing compilers for today's embedded processors is faced with a wealth of compilation techniques designed initially for architectures of a wide variety. These techniques have been converging from two main areas: software compilation for general-purpose microprocessors and high-level synthesis for ASICs [36]. This chapter presents a review of individual techniques with emphasis on the methods which apply to the constraints imposed by today's embedded processors.

2.1 Traditional software compilation

This section discusses the well-known techniques used in most compilers for general purpose computing systems such as workstations and personal computers. The content is restricted to characteristics of the compilation problem as they pertain to real-time embedded systems. In such a system, code performance and size is critical, since the firmware is intended to reside in the system reacting only to external stimuli. The effectiveness of a compiler is of utmost importance, since for embedded processors, it may mean the difference between the compiler being used and not!

2.1.1 Dragon-book compilation

The classic text by Aho, Sehti, and Ullman [1] defines compilation as the translation of a program in a source language (e.g. C) to the equivalent program in a target language (e.g. assembly code and absolute machine code). This translation is typically decomposed into a series of phases, as shown in Figure 2.1.

The first two phases of the process deal with parsing the physical tokens of the source program (lexical analysis) and analyzing the structure of the programming language (syntax analysis). The result of this is an intermediate representation of the source code. A typical example of this representation is a forest of syntax trees (Figure 2.1). For each tree, a node represents either an operation (e.g. =, +) which is to be executed upon its children nodes, or the identifier of a symbol.

The third phase is an analysis of the intended meaning of the language (semantic

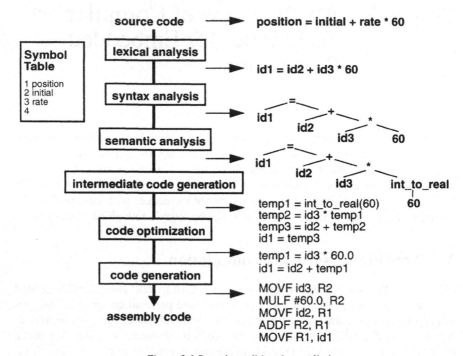

Figure 2.1 Steps in traditional compilation

analysis). It statically determines that the semantic conventions of the source language are not violated. Examples of semantic checks are: type checking, flow-of-control checking, and symbol name checking.

These three phases need not be totally independent, but are often sequentially executed *on-the-fly*, during the traversal of the source program. In tandem with these three processes, one or more symbol tables are constructed as an internal housekeeping of the compiler for symbol types, sizes, locations, etc.

Following these phases, many compilers produce an intermediate code, which can be thought of as code for an abstract or *virtual* machine. A common form is known as *three-address code* (or *tuples*), which simply means that each instruction has at most three operands: 2 sources and 1 destination.

This intermediate code can be improved upon using code optimizations of which a large number and variety exist [1][28]. These range from local to global optimizations and from guaranteed improvements to high gain, high risk transformations. That could mean that after an optimization, code is worse than the original in terms of area and performance. Choosing the right level of optimization is a difficult task; however, it cannot be disregarded. It is often the case that the result of compilation is unsatisfactory without the application of optimizations. Optimizations for embedded processors are discussed in Section 2.6.

Finally, code from the intermediate form is translated to assembly code for the target. Memory locations are chosen for variables (register and memory allocation) and the code that results is suitable to be run on the target machine.

In the context of compilers for embedded processors, there are clearly some difficulties with this traditional approach to software compilation. We outline some of the main issues as follows:

1. Retargetability. In the traditional approach to compilation, retargeting to a new architecture is confined to the final code generation phase. This means that the intermediate code must closely resemble the final target in order to produce efficient code. If the instruction-set of the final target is widely different than that of the virtual machine, it can be difficult to produce efficient target code. As embedded processor instruction-sets vary widely in composition, it may be troublesome to conceptualize an intermediate form which is general enough for any target.

2. Register Constraints. Embedded processors often contain a number of special-purpose registers as opposed to general purpose register files. In many cases, registers are reserved for special functions. This is a design effort used to narrow instruction words through format encoding. The instruction width reflects directly into program space, which is costly especially for on-chip programs. The impact of register constraints is on all the phases of compilation.

3. Arithmetic Specialization. Three-address code artificially decomposes data-flow operations into small pieces. Arithmetic operations which require more than three operands are not naturally handled with three-address code. Operations such as these often occur on DSP architectures [64].

4. Instruction-level Parallelism. The task decomposition in the traditional view of compilation does not naturally suit architectures with parallel executing engines. For example, a DSP often has both data calculation units (DCU) and address calculation units (ACU). A compiler should take into account the possibility to perform operations on different functional units, as well as choose the most compact solution.

5. Optimizations. Real-time embedded firmware cannot afford to have performance penalties as a result of poor compilation. Efficient compilation is only arrived upon by many optimization algorithms. Optimizations to intermediate code (e.g. three-address code) are mainly restricted to a local scope. Global optimizations which use data-structures (e.g. arrays, structures), data-flow, and control-flow information would be more naturally suited to a higher level intermediate representation, closer to the source program structure. Even at this high level, optimizations should take into account the characteristics of the target architecture.

Many techniques are beginning to be introduced to overcome these and other factors which can make compilation for embedded processors much different than for general computing architectures. Some of the new approaches improve on weaknesses of the traditional view of compilation, while others introduce new methods which have evolved from techniques in behavioral and register-transfer level hardware synthesis.

2.1.2 The GNU gcc compiler

The GNU gcc compiler is distributed by the Free Software Foundation [96] and originates by work of Richard Stallman, Jack Davidson, and Christopher Fraser, with contributions by many others. With the free distribution of its C source code, it has been ported to countless machines and has been retargeted to even more. For examples of embedded systems, a retargeted gcc compiler is offered commercially for several DSPs including the Analog Devices 2101, the AT&T 1610, the Motorola 56001, the SGS-Thomson D950, and the DSP Group Pine and Oak cores. It has become the de-facto approach to develop compilers quickly from freely available sources.

A simplified picture of the gcc compilation flow is shown in Figure 2.2. The input source code is parsed and converted into an internal form, called Register Transfer Language (RTL), inspired by LISP lists. A number of architecture independent optimizations are applied to the RTL prior to any further transformation. The optimized RTL is then refined during the following phases: instruction combining groups simple RTL operations into clusters of operations; instruction scheduling orders the instructions in the time axis (see Section 2.5 on scheduling); register class

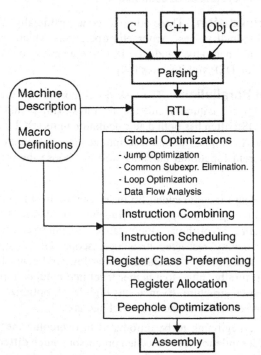

Figure 2.2 A simplified view of the GNU gcc compilation chain

referencing selects the most appropriate register file for each live variable, while registers within register files are allocated and assigned during register allocation (see Section 2.4 on register allocation and assignment); a final machine-specific peephole optimization phase is applied to the generated code (see Section 2.6.1 on peephole optimization). Several of the above phases depend on the machine description, mostly for what concerns the available instruction patterns and peephole optimizations. Machine-specific macros and functions written in C are also used in the machine description.

One of the well-known strengths of gcc is its set of architecture *independent* optimizations: common subexpression elimination, dead code removal, constant folding, constant propagation, basic code motion, and other classical optimizations. However, for embedded processors, it is extremely important that optimizations be applied according to the characteristics of the target architectures. Simple and often *innocent-looking* optimizations can have adverse effects on the efficiency of code. This is discussed in Section 2.6. Unfortunately, gcc has little provisions for which optimizations may be applied according to the target machine.

In the area of real-time DSP systems, the performance of many of the gcc-based compilers fall short of producing acceptable code quality. This has been demonstrated clearly by the DSPStone benchmarking activities [112][113]. Further evaluation of DSP tools and compilers for commercial processors have been done by Berkeley Design Technology Inc. [13] which also show that many of the existing commercial C compilers produce inefficient code on commercial fixed-point DSPs [15].

This result may be somewhat surprising given that this compiler technology has existed for some time. In particular, gcc is very popular as an efficient compiler for many workstation and home computing systems. The underlying reason for the difference in performance between general computing targets and DSPs is made clear in the document distributed with gcc. Quoting from [96]:

"The main goal of GNU CC was to make a good, fast compiler for machines in the class that the GNU system aims to run on: 32-bit machines that address 8-bit bytes and have several general registers. Elegance, theoretical power and simplicity are only secondary. GNU CC gets most of the information about the target machine from a machine description which gives an algebraic formula for each of the machine's instructions. This is a very clean way to describe the target. But when the compiler needs information that is difficult to express in this fashion, I have not hesitated to define an ad-hoc parameter to the machine description. The purpose of portability is to reduce the total work needed on the compiler; it was not of interest for its own sake."

Embedded processors usually fall into the category of having few registers, heterogeneous register structures, unusual word-lengths, and other architectural specializations. gcc was not conceived for these types of processors; and therefore, compiler developers using gcc are faced with two choices: lower code quality or a significant investment in custom optimization and mappings to the architecture. In the latter case, *ad-hoc* parameters in the machine description and machine-specific routines are needed. Naturally, this greatly reduces the compiler retargetability.

2.2 Compiler retargetability

Ever since the appearance of compiler technology, an interest in retargetability was raised to support the varying architecture design styles and also to support processor upgrades [32]. While there was always interest in the topic, a formal retargetability model has never been fully adopted. The trend has been that the more optimization effort put into a compiler the more that compiler becomes invariably linked with the specific architecture. Furthermore, the lifetime of an instruction-set for a general computing processor has usually been long enough to justify concentrating all the effort on architecture-specific compilation.

For embedded processors, the renewed interest in retargetable compilers is two-fold:

1. Retargetability allows the rapid set-up of a compiler to a newly designed processor. This can be an enormous boost for algorithm developers wishing to evaluate the efficiency of application code on different existing architectures.

2. Retargetability permits architecture exploration. The processor designer is able to tune his/her architecture to run efficiently for a set of source applications in a particular domain, recompiling the application for each redesign of the architecture.

Ideally, a truly retargetable compiler is one whereby the programmer himself is able to reconfigure the compiler simply by changing the specification of the compiler. The principle is shown in Figure 2.3.

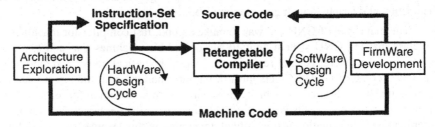

Figure 2.3 The retargetable compilation principle

Figure 2.3 shows two design cycles: the software and the hardware design cycle. The one to the right of the figure is the familiar development course, where the programmer uses the compiler to develop software. The second cycle is to the left of the figure, showing the retargetable compiler being used as a design tool to explore the processor architecture. The ideal user-retargetable situation is where the instruction-set specification completely describes the processor mechanics in a manner which is simple enough so that the programmer is able to make changes himself. Exploration is supported for redesigns by changes to the instruction-set specification.

2.2.1 Different levels of retargetability

For today's compilers a great many levels of retargetability exist. In [6], Araujo classifies retargetability into three categories. A general interpretation of these categories is as follows:

1. Automatically retargetable: the compiler contains a set of well defined parameters which allow complete retargeting to the new processor. Full knowledge of the range of target architectures is contained within the compiler. Retargeting time is on the order of minutes and seconds.

2. User retargetable: the compiler user is able to retarget the processor by furnishing an instruction-set specification. The compiler may require a certain amount of pre-compile or set-up time. Retargeting time is on the order of days and hours.

3. Developer retargetable: the compiler may be retargeted to a range of processor architectures, but requires expertise with the compiler system. This category can become blurred with the complete rewriting of a new compiler. Retargeting time is on the order of months and weeks.

While the dividing line between these categories can be difficult to place, perhaps the most indicative measure is the retargeting time, which clearly separates the classes.

State-of-the-art compilers for embedded processors fall primarily in categories 1 and 3, while the main goal is to fall into category 2. Compilers in category 1 are mainly single-target compilers which allow small variations to the target processor. The weakness in these compilers is the small range of targets which they support, therefore making architecture exploration difficult.

The advantage of compilers in category 3 is the support for a wide range of architectures. However, the weakness is the relatively long compiler development time. In addition, a compiler expert is needed to perform the compiler retargeting.

2.2.2 Architecture specification languages and models

The most promising avenue for supporting the retargetability of compilers for embedded processors is the work on specification languages and models. An instruction-set specification language allows a user to describe the functionality of a processor in a formal fashion. Subsequently, the transformations of a compiler may be retuned according to the architecture. This retuning can be done by means of an architecture model. While an architecture model need not be generated by a specification language, a language-based input is the most natural interface to the user.

Mimola. The MSSV/Q compilers from University of Dortmund [73][74] represent early work on mapping high-level algorithms to structural representations of processors. The processor is described in a hardware description language called *Mimola* [11]. The structure of the processor is defined by a netlist of functional components with the explicit activation of components via bits in the instruction word. An example is shown in Figure 2.4. This example target Mimola structure contains a dual-port

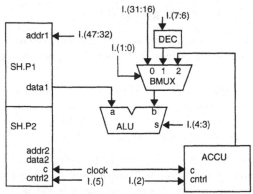

Figure 2.4 An example Mimola target structure [74]

memory (SH), an Arithmetic and Logic Unit (ALU), an accumulator register (ACCU), a decoder (DEC), and a multiplexer (BMUX). Sets of bits from the instruction word (I) activate functional units and connections in the target architecture.

The algorithm language is a Pascal-like subset of Mimola. After the application of a set of target-dependent, user-definable program transformation rules, the algorithm is matched to the target structure. The compiler uses a recursive descent algorithm matching operations to functional units, and constants and variables to memory locations. During this execution, paths are matched using reachability analysis of the target structure. For optimization reasons, several instruction versions are generated and bundling is performed to reduce the number of final instructions.

The strength of the MSSQ/V compilers is the direct description of the processor architecture. The compilers work directly with the physical structure of the hardware, which leads to a fair level of retargetability. However, writing the processor description requires intimate detail of the decoding strategy of the entire architecture. In the case of commercial processors, for example, detailed information of the hardware would not be available. Only a programmers manual of the instruction-set is available. Furthermore, matching an algorithm to a detailed netlist of components could reflect in inefficient compilation times as physical paths need to be frequently traced.

The more recent generation of compilers from University of Dortmund is the Record [60][61] compiler. It also uses Mimola as the processor description language but uses another approach for compilation. Record uses a pre-compilation phase called *instruction-set extraction* which automatically generates a compiler code selector from a hardware description model of the processor. The advantage of this approach is the use of an efficient code generator generator (this concept is discussed in Section 2.3.1) for the main instruction-set selection phase of compilation. It is this improvement that separates Record from the MSSQ/V compilers. In addition to this are a number of additional compiler transformations which improve instruction-level parallelism, including address generation (discussed in Section 5.2.1) and compaction (discussed in Section 2.5).

The instruction extraction procedure begins from a Mimola netlist model of the processor and by traversing the data paths, it determines the instructions which may be executed on the processor. These patterns are then fed to a program which produces a grammar for the code generator generator. In this manner, a set of register-transfer templates are formed which, when coupled with the code generator, comprise the front-end of the compiler.

The Record approach is indeed a large improvement over the previous MSSQ/V compilers as it makes use of structural information of the processor while allowing efficient pattern matching utilities to be used for the main compilation flow. Again as the Mimola hardware description language is used, the disadvantage of the approach is that an explicit netlist of the processor is needed to retarget the compiler. While this might be an advantage to a hardware designer who deals with his own specialized embedded processors, it is disadvantageous for a programmer using a commercial processor where the details of the hardware functionality are unknown.

nML. The CBC compiler [24][25] from the Technical University of Berlin is a project that inspired the development of a processor specification language known as nML [31], which stands for *not a Machine Language*. This inventive abbreviation stresses the fact that the language is intended to describe the behavior of a processor rather than the structural details. The nML language describes a processor by means of the instruction-set and the execution mechanics of that instruction-set. The key elements of the language are the description of operations, storage elements, binary and assembly syntax, and an execution model. These elements combined with some features such as the derivation of attributes allows the full description of an instruction-set processor without the detailed structural information of a netlist. The level of information is comparable to a programmer's manual for the processor.

The nML language is based on a synchronous register-transfer model, allowing also the description of detailed timing including structural pipelining. Figure 2.5 shows a partial instruction-set description demonstrating the principal elements of the nML language. Types may be composed from a set of pre-defined type constructors: `bool`, `card`, `int`, `fix` and `float`. Using these types, storage elements may be declared while providing names for identification. The last principal element of the language is the *partial instruction* (PI), which is described in one of two ways: an *OR-rule* which declares several alternatives; and an *AND-rule* which combines several PIs to form a new PI. This is done in attribute grammar, whereby each PI may be derived from other PIs. This is shown conceptually in Figure 2.6.

Associated with each PI is a continued set of attributes, the foremost being the `action` attribute which describes the execution behavior. The `action` attribute may be of two permitted forms:

1. assignment: e.g. `dst = src1 + src2`

2. conditional: e.g. `if c then dst = 0 else dst = 1 end`

Declarations

Storage Elements

Partial Instructions

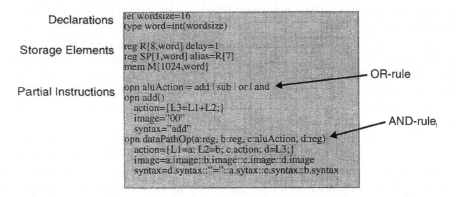

Figure 2.5 Sample nML Language Elements

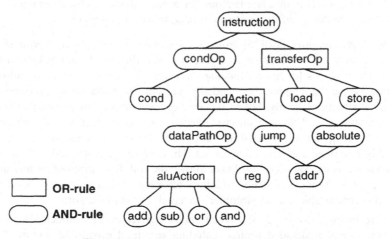

Figure 2.6 The derivation of attributes in the nML language.

The definition of the behavior may contain any operators from a pre-defined list, including C-like operators: arithmetic & logic, bit rotation operators, and type conversion functions.

Other important attributes of the PI are `image` and `syntax`, which describe both the binary and assembly representation of the microcode. The compiler uses these attributes to determine the fashion in which to emit the microcode.

The nML description language is a complete description of the processor at the level of a programmer's manual. For embedded processors, the user is able to capture all the functionality, execution, and encoding of the machine. The strong point of nML is that the language is not tied to the implementation of a compiler or simulator, which is the case for many machine descriptions [96].

Instruction Set Graph. The ISG model was introduced by VanPraet [106] and is used in the Chess compiler [57]. The representation is an example of a model that associates behavioral information of the processor with structural information. Making use of nML as the description language, the ISG model is generated automatically and encapsulates the functionality of the processor together with the instruction-level semantics. The main elements of the Instruction Set Graph are shown in Figure 2.7. The ISG contains two types of storage elements: *static resources* such as addressable memory or registers with explicit bit-widths, and *transitories* which pass values with no corresponding delay. Storage elements are interconnected by *micro-operations* which correspond to specific operations which may be executed on a functional unit of the processor by an instruction code, or by connectivity to other storage elements. Each micro-operation contains a list of legal instruction-bit settings, which in principle activate a connection between storage elements. Therefore, a legal micro-instruction is constructed by forming a path through the ISG, keeping structural hazards in mind.

This approach allows a convenient encapsulation of the operations of the processor while keeping an active record of the encoding restrictions defined by the instruction-set. In the Chess compiler it serves as a base model for all the phases of compilation (instruction-set matching and selection, register allocation and assignment, and scheduling) to form the mapping from a source algorithm to microcode

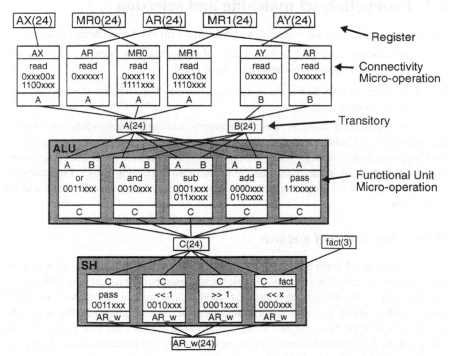

Figure 2.7 Principles of the Chess Instruction-Set Graph (ISG) model

implementation. The model is a higher abstraction than a full functional unit netlist which ties behavioral operation to the structural data connectivity.

CodeSyn Model. The *CodeSyn* compiler developed at Bell-Northern Research / Nortel uses a mixed structural and behavioral level model to describe the target instruction-set processor. Similarly to the ISG model, it ties behavioral aspects with the structure of the architecture. Details of the model are described in Section 3.2.2.

Two main goals remain for compilers targeting embedded processors: code quality and retargetability. While the goals are sometimes conflicting, the growing amount of embedded firmware and the rapid appearance of new architectures makes both equally important. For these goals, there are three principal compiler tasks for embedded processors:

- instruction-set matching and selection,
- register allocation and assignment,
- scheduling and compaction.

Unfortunately, the three tasks are highly interdependent, which is a concept known in the compiler community as *phase coupling*.

2.3   Instruction-set matching and selection

We separate instruction-set matching and selection into two broad definitions:

1. Instruction-set matching is the process of determining a wide set of target instructions which can implement the source code.

2. Instruction-set selection is the process of choosing the best subset of instructions from the matched set.

While these general definitions could be interpreted as the entire compilation process, the matching and selection process has varying levels of importance depending on the compilation approach. In some compilers, it does comprise the entire compilation process; in others, it is only one phase of other more important phases. Furthermore, some compilers take a simplified view of the process, selecting only the first matched instructions.

2.3.1 Pattern-based methods

The traditional approach to matching source code to an instruction-set is to produce a base of template patterns of which each member represents an instruction. During compilation of a source program, these patterns are matched to portions of the source. For example, it is possible to translate a source program into a forest of syntax trees, which are then matched to the pattern set of syntax trees (see Figure 2.1). A subset of all the matched patterns are selected to form the implementation in microcode (i.e. instruction-set selection).

Dynamic programming [3] is a method used to select a *cover* of patterns for the subject tree. It is a procedure with linear complexity that selects an optimal set of patterns when restricting the problem to trees and a homogeneous register set [3]. The procedure is a simple linear process which guarantees the best choice of patterns at each node of a tree (the procedure is described in [1]). However, embedded processors are characterized by heterogeneous register sets and instructions best described by graph-based patterns. This means the advantage of dynamic programming is diminished for embedded processors.

Tree-based pattern selection extensions which allow the handling of heterogeneous register sets have been formulated in the work of Wess [108][109]. In this approach, register constraints are encapsulated by a *trellis* diagram. Using this diagram as the target model, the code selection process is considered as a path minimization problem.

On the level of software engineering, a popular, and interesting approach is the so-called *code generator generator* or *compiler compiler*. Examples of these systems are the Glanville-Graham generators [35], BEG [23], Twig [3], Burg and Iburg [124][30], and Olive [123]. We present the concepts in two steps: code generation by tree rewriting, and pattern matching by parsing [1].

Tree rewriting. A simple example is shown in Figure 2.8, where the source code is represented by a syntax tree. A set of reducing rules allows the tree to be rewritten by successive applications. For each application of a reducing rule to a branch of the tree, code is emitted. In the example, the rules are applied in the following order: 2, 1, 3. Although this is a simple example, it illustrates the procedure which applies to trees of any size and shape and can also be extended to dags [28].

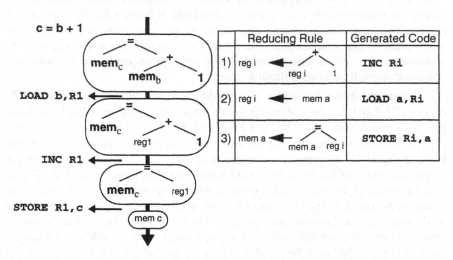

Figure 2.8 Principle of code generation by tree rewriting.

Pattern matching by parsing. It was observed in [35] that the matching of code templates against an expression tree resembles the problem of matching productions against a token sequence during source code parsing. Representing the syntax tree as a prefix string allows the transformation of the previous problem into a parsing problem. For example, the syntax tree of Figure 2.8 becomes:

```
= memc + memb 1
```

In this way the reducing rules become simply a grammar with related actions:

```
1) regi <- + regi 1 {INC Ri}

2) regi <- mema {LOAD a,Ri}

3) mema <- + mema reg i {STORE Ri,a}
```

Three main issues arise with the code generator generation principle:

1. when more than one pattern matches a tree (i.e. conflicts), the quality of code is dependent on which rule is applied. (i.e. pattern size trade-offs)

2. the quality of the code is dependent on which branches of the tree are visited first (i.e. scheduling)

3. registers are chosen *on-the-fly* (i.e. register assignment is local)

The first of these can be approached by simply favoring larger patterns; however, this is an ad-hoc approach which does not always reflect the cost of a pattern. It is possible to use dynamic programming in this stage [3], even at compile-compile time when based on homogeneous registers and a constant cost model [30]. However, it may be a disadvantage when incorporating more complicated cost models which depend on heterogeneous registers and pattern selections crossing tree boundaries. Scheduling and register assignment are extremely important issues for embedded processors requiring code quality and are discussed in Section 2.5 and Section 2.4 respectively. In general, the disadvantage of a code generator generator is that it integrates many of the compiler phases into one. Consequently, when it is important to concentrate on a certain phase of compilation which is important for an embedded processor target, it becomes difficult to tackle.

Despite the basic difficulties, the SPAM (Synopsys Princeton Aachen MIT) project [7][62][63] has been able to apply the principles to one embedded processor, the Texas Instruments TMS320C25 DSP. Using the Olive code generator generator [123], a grammar was constructed for the TI C25 data-path shown in Figure 2.9. The Olive grammar for the architecture is shown in Figure 2.10. The use of a grammar allows the details of the parser to be hidden from the user; however, during this process a schedule of instructions must be considered before code is emitted. This is discussed in Section 2.5. The Olive pattern set treats the restrictions of registers for the TI C25 architecture; however, it is not clear whether it is possible to specify processors with larger register files with overlapping register roles. In addition, the restrictions of register uses with the encoding of the instruction-set is not included in the grammar.

Other approaches which use tree-based pattern matching and selection include

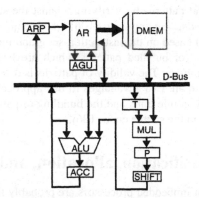

Figure 2.9 The Texas Instruments TMS320 C25 data-path.

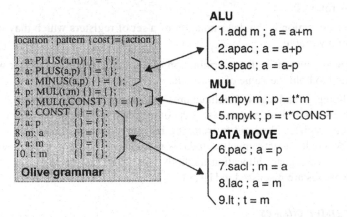

Figure 2.10 SPAM Olive grammar and patterns for the TI C25.

the *CodeSyn* compiler and *FlexCC* compilers, which are presented in Chapter 4.

2.3.2 Constructive methods

Although not yet extensively studied for embedded processors, techniques for the matching of directed acyclic graphs (*dags*) as opposed to simple tree structures may become important. This is because many instructions in embedded processors and DSPs are more naturally described as dags, for example: accumulator-based machines and auto-increment (decrement) address registers. In addition, more optimization possibilities are available from source algorithms when all the control and data-dependencies are explicitly kept in a dag. As generating code for trees is a difficult problem, generating code for dags becomes much more complicated. Heuristics which enhance tree-based methods are explained in [28] and [1].

Some approaches have been introduced which base pattern matches on structural connections of the processor. Using the Mimola model as described in Section 2.2.2,

MSSQ/V determine valid patterns by verifying against the structure. Similarly, the Chess compiler uses a *bundling* approach [106] which couples nodes of a control-data flow graph (CDFG) based on the instruction-set graph model (Figure 2.7). This also allows the selection of bundled patterns which are heavily restricted by the encoding of the instruction-set. The validity of patterns is determined directly by the architecture model. There are two advantages to this approach: the pattern set need not be computed at pre-compile time, and the bundling of patterns can possibly pass control-flow boundaries in the source code [106].

2.4 Register classification, allocation, and assignment

Issues in data storage for embedded processors are probably the most difficult problem in compilation. The largest tasks are register allocation and assignment, which we define as follows:

1. Register allocation is the determination of a set of registers which may hold the value of a variable.

2. Register assignment is the determination of a specific physical register which is stipulated to hold the value of a variable.

As well as being a challenge for architectures with many registers, register allocation and assignment for embedded processors is complicated by special-purpose registers, heterogeneous register files, and overlapping register functions. However, as a large base of work already exists, much of today's research builds upon techniques of the past. A survey of register allocation methods until 1984 can be found in [79], and classic approaches are discussed in [1][28].

2.4.1 Register classes

In dealing with heterogeneous register files of programmable machines, one approach is to introduce the concept of *register classes* [27][96]. In general, a register may belong to one or more register classes of overlapping functionality. By these means, the compiler is able to calculate those registers which are most needed for a specific function; and hence, a strategy for register allocation and assignment can be carried out.

In both the CBC compiler [27] and the GNU gcc compiler [96], the concept of a *symbolic* or *pseudo* register is used so that the compilation may proceed in two steps. During instruction-set selection, a symbolic register is assigned for each program storage element. Next, symbolic registers are organized by means of the register classes in a register allocation phase. Following, detailed register assignment of each symbolic register to a real (i.e. *physical, hard*) register is performed.

The *CodeSyn* compiler builds upon the concept of register classes for special-purpose registers and the allocation approach is described in Section 3.2.4.

2.4.2 Coloring approaches

Pioneering work on register allocation by Chaitin [17][18] introduced the notion of *coloring* to determine the number of registers needed for a program's variables. An example is shown in Figure 2.11 to explain the coloring formulation. The left part of

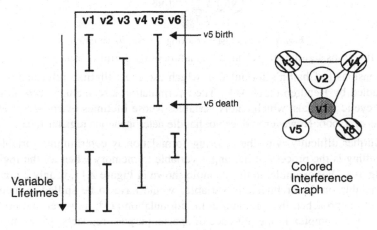

Figure 2.11 Register allocation by coloring.

Figure 2.11 shows a basic block of source code which contains a set of variables and their corresponding lifetimes. A variable is said to be alive when it's value must be retained for an operation which occurs later in the program. The process of coloring proceeds in two steps. The first step is to build an *interference graph* whereby the nodes of the graph represent the variables and a set of edges connecting the nodes. An edge represents an overlap of lifetimes between the two variables, meaning that the two variables cannot use the same storage unit (register).

The second step is to assign a color to each node of the interference graph, such that no two connecting nodes have the same color. The number of colors used in the graph is the number of registers needed in the program. The interference graph shown in Figure 2.11 needs at least 4 different colors and therefore 4 registers.

Taking a closer look at this example, one would notice that there is solution to this problem which uses just 3 registers. The difficulty was in the previous formulation of the coloring problem. That interference graph has its weaknesses, as it contains neither the overlapping information of lifetimes nor the relative times of the overlaps. To arrive at the 3 register solution, the interference graph in Figure 2.12 should be solved. The variable v1, which has two independent lifetimes has been split into two nodes, allowing it to reside in two different registers.

In a real program, the coloring formulation is entangled with control-flow constructs such as if-then-else conditionals, loops, case statements, function calls and local/global scoping. The formulation using coloring for a real application must be extended to handle these cases. In addition, heterogeneous register files and over-

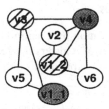

Figure 2.12 Register coloring: second formulation

lapping functions significantly change the nature of the formulation.

Extensions to Chaitin's formulation which use exact lifetime information have been studied by Hendren et. al. [41]. Their formulation also includes careful treatment of cyclic intervals, which colors variables whose lifetimes extend over several iterations of a loop. However, it does not handle heterogeneous register files.

A further difficulty with the coloring formulation is determining variables to *spill*. Spilling is the process of moving a variable to memory when all the registers are being used. For example, in the example shown in Figure 2.11, if only 2 registers were available on the architecture, variables would have to be spilled to memory. Chaitin has approached this problem in his formulation [18]; however, the problem becomes more complex in the presence of special purpose registers; for example, if only certain registers are permitted to store a value to memory (see Section 6.1).

Another formulation of the register allocation problem, which can be regarded as a type of coloring, is inspired by the channel routing problem in place and route synthesis. The *Left Edge* algorithm [54], which has been used for channel routing of connections in VLSI physical design, can also be applied to register allocation. In Figure 2.13 we illustrate the formulation using the previous example.

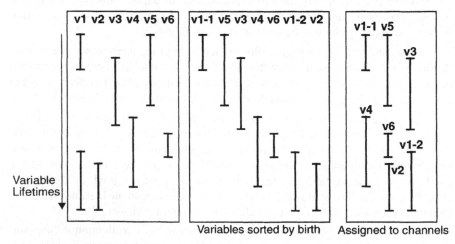

Figure 2.13 The Left Edge algorithm applied to register allocation.

The approach proceeds in four steps:

1. Sort the variable lifetimes (i.e. segment) in increasing order of their births or left-edge (top-edge in the case of Figure 2.13).

2. Assign the first segment (top-most edge) to the first channel. (e.g. v1-1 in Figure 2.13)

3. Find the next segment which can be placed in the current channel (e.g. v4 in Figure 2.13).

4. Continue until no more segments fit into the current channel. Start a new channel and repeat Step 2. until all segments are assigned to channels. When finished, each channel may be assigned to a physical register.

The Left Edge algorithm is greedy in nature; however, it produces an optimal result in the minimum number of registers needed. Notice that the example of Figure 2.13 requires a minimum of 3 registers as was discovered with the coloring formulation in Figure 2.12. The advantage of the Left Edge algorithm over coloring is that it explicitly takes register lifetimes into account. Naturally, the approach must be extended to take into account control structures, spilling, and special-purpose register structures [66].

Coloring is a problem formulation to register allocation. After colors are determined, the register assignment part is a simple one-to-one mapping of colors (or tracks) to registers. Register assignment is not so simple in the presence of special-purpose registers; however, a solution can be provided using register classes as described in Section 3.2.4.

2.4.3 Data-routing

To deal with the distributed nature of registers in DSPs and application-specific signal processors, the register assignment process has been reformulated as a problem closely tied to the architecture structure. In *data-routing*, the goal is to determine the best flow of data through the architecture such that execution time of the microcode is minimized. While data-routing techniques are generally more time consuming than other register assignment approaches, they usually provide solutions for architectures with very heavily constrained register resources and distributed register connections. Some of the previously mentioned register allocation techniques can fail for these architectures.

Rimey and Hilfinger [90] introduced an approach known as *lazy data-routing* for generating compact code for architectures with unusual pipeline topologies. The idea is to schedule instructions as compact as possible and to decide on a data-route only after an operation is scheduled. A spill path to memory is always guarded to guarantee that no deadlocks will occur. A similar approach was used by Hartmann [40] in the CBC compiler, where a complicated deadlock avoidance routine was incorporated for architectures with very few registers.

The Chess compiler uses a branch and bound approach to data-routing [56][57].

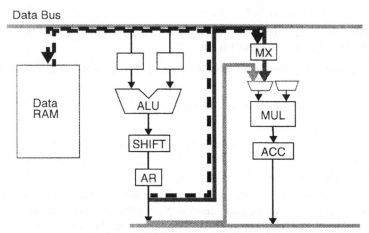

Figure 2.14 Data-routing in the Chess compiler.

An example of the concept is shown in Figure 2.14. Three alternate data-routes are shown for a value which leaves the AR register and is headed for a multiplication. The first is the direct route along the lower bus and directly into the multiplier (MUL), the second uses a longer path by means of the upper bus and the input register of the multiplier (MX), and the third shows the value temporarily spilled to data memory (DATA RAM). The data-routing approach determines a solution from these three candidates which were found through branch and bound search techniques. The quality of each solution is determined using probabilistic estimators which monitor the impact of an assignment on the overall schedule of the control-data flow graph of the source code. These estimators are based on scheduling algorithms from high-level synthesis [85].

2.5 Scheduling and compaction

Scheduling is the process of determining an order of execution of instructions. Although it can be treated separately, the interdependence with instruction selection and register allocation makes it a particularly difficult problem for embedded processors. Furthermore, machines which support instruction-level parallelism require fine-grained scheduling. This type of scheduling is called *compaction*.

An example shown in Figure 2.15 for the Texas Instruments TMS320C25 has motivated scheduling techniques in the SPAM project [7]. Recalling the instruction-set patterns shown in Figure 2.10, this example shows a data-flow tree which has been scheduled three different ways. The *Normal Form Schedule* [2] can generate optimal code for architectures with homogeneous register sets, but the example in Figure 2.15 a) shows that suboptimal results can occur for the TI C25 even for a very simple example. Problems stem from the very few and distributed registers of the architecture (see Figure 2.9). Only a clever schedule can avoid the need to spill and reload values from extra memory locations as shown in Figure 2.15 b) and Figure 2.16 a). The optimal solution (Figure 2.16 b) requires consideration of the architecture's register and memory structure.

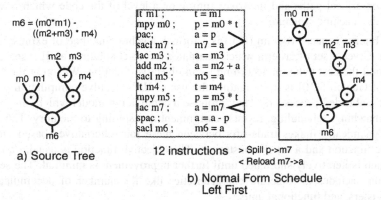

12 instructions   > Spill p->m7
< Reload m7->a

b) Normal Form Schedule
Left First

Figure 2.15 Motivating data-flow example for the SPAM project with the TMS320C25 [7]

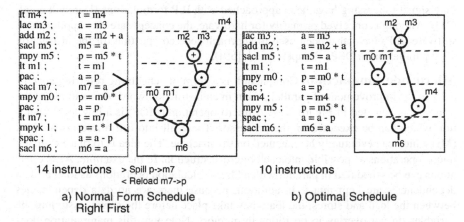

14 instructions   > Spill p->m7
< Reload m7->p

a) Normal Form Schedule
Right First

10 instructions

b) Optimal Schedule

Figure 2.16 SPAM scheduling example for the TMS320C25

The SPAM group have proposed a solution which optimally solves tree-based data-flow schedules for architectures such as the TI C25 architecture, which satisfy certain criteria. These criteria include the presence of only single and infinite storage resources connected in a certain way. Extensions would be necessary for the register structures of architectures which differ substantially from the storage criteria displayed by the TI C25. Furthermore, the approach neither handles control-flow in the source code, nor instruction-level parallelism of the processor.

Mutation scheduling [81] is an approach whereby different implementations of instructions can be regenerated by means of a mutation set. After the generation of three-address code, critical paths are calculated. Attempts are made to improve the speed by identifying the instructions which lie on critical paths and *mutating* them to other implementations which allow a rescheduling of the instructions. In this manner, the overall schedule is improved. The advantage of this approach is that it works directly on critical paths and improves timing on a level of the code which is very close to the machine structure.

An Integer Linear Program (ILP) is a formal, algebraic method of expressing a problem. Given a set of criteria which guarantee a correct solution, an ILP solver is able to find the best solution according to an objective function. Wilson at the University of Guelph [110] is investigating the use of ILPs to solve compilation problems for embedded processors. They propose a compilation model which integrates pattern-matching, scheduling, register assignment and spilling to memory. The ILP solver dynamically makes trade-offs between these four alternatives based on an objective function and a set of constraints. The objective function is usually a time goal which is iteratively shortened until further improvement is minimal. The set of constraints includes architecture characteristics like the number of accumulators, other registers, and functional units.

Scheduling and also software pipelining (discussed in Section 2.6.2) for real-time signal processing have been approached with ILP formulations (Depuydt et. al. [21]). Although conceived primarily for hardware, the concepts are also applicable to software. ILPs have also been used to approach code compaction for the instruction level parallelism in DSPs (Leupers et. al. [59]).

Microcode compaction. Compaction is a form of scheduling referring specifically to the improvement of parallelization in an instruction word. An example of the principle is depicted in Figure 2.17. A micro-operation (MOP) is a low-level operation which can be executed on the processor. These fill into full micro-instructions (MI) which can eventually be executed on the machine. The idea is to place as many micro-operations as possible into each micro-instruction. In the example, ALU operations can be scheduled in parallel with address calculation operations as long as data dependencies are kept intact. In addition, because there are no data dependencies between the load and stores, the load may take place before the stores, assuming the variables do not overlay in positions in memory. Note that this manipulation does increase the lifetime of register R1; however, if this scheduling is done on the level of micro-instructions, register assignment has already been done and the compaction will be inhibited in cases where the register data-dependence is violated. For example, this compaction would not be possible if register R4 were used for the load instruction.

Lioy and Mezzalama [71] have approached the compaction problem by defining pseudo micro-instructions and sequences of micro-operations with source and destination properties. These sequences can then be packed into and upward past pseudo micro-instructions to form real micro-instructions. This packing takes into account the resource conflicts of the machine, such as register dependencies and the use of functional units.

For embedded processors, the compaction problem is more intricate because of

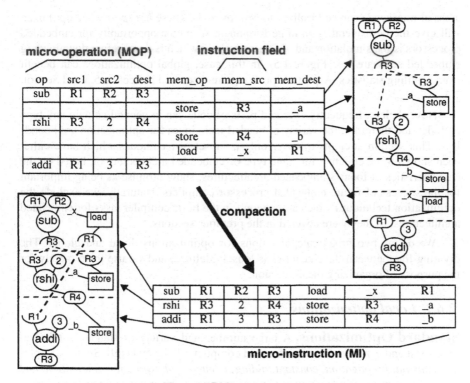

Figure 2.17 Microcode compaction.

the possible encoding restrictions of the architecture as explained in Section 1.3.2. Parallelism in a machine architecture does not always imply that the instruction-word supports that parallelism. Again, this stresses the point of phase coupling with the other tasks of compilation like instruction-set selection and register allocation. For microcode compaction, a highly encoded instruction-word also means that instruction bit-fields cannot be simply regarded as an orthogonal resource.

2.6 Optimizations for embedded processors

The subject of compiler optimization for embedded processors remains predominantly an open problem. While a large amount of optimization theory exists for general computing architectures [1][9][10][28], the topic is not well understood for embedded real-time architectures. This is primarily because the standard mapping techniques for embedded architectures are just beginning to appear [75]; and secondly, because of the amplitude of constraints that embedded processors impose on standard optimization techniques.

For real-time embedded processors, the *rule of thumb* is the 90/10 rule. The code spends 90% of the time in 10% of the code. This simply emphasizes that time-critical areas of microcode can be localized to certain areas. These *hot spots* when optimized will give the best overall gain in performance. A unique opportunity for embedded processors is the simulation and verification cycle which is done before downloading embedded software (see Figure 1.5). In this case, global optimizations can benefit from profiling statistics. Aspects of profiling are discussed in Section 5.3 and Section 7.5.

Nevertheless, frequently executed parts of code can in general contain any type of code; and therefore, the variety of needed techniques for optimization are boundless. This section attempts to cover only a subset of techniques which have either been shown to be effective for some type of embedded architecture or, based on the characteristics of today's embedded architectures, show promise as being important for the success of future embedded processor compilers. Naturally, we exclude the optimization techniques which are inherent in the basic compiler tasks for embedded architectures, as they were covered in the previous sections.

We define two broad categorizations for optimizations: local and global. The dividing line between the categories is loosely defined, and we use them merely for the purpose of organizing the discussion.

2.6.1 Local optimizations

Standard Optimizations. A large number and variety of standard optimizations exist and can have various effects on compiler efficiency [1][9]. Some examples are: *constant propagation*, *constant folding*, *common subexpression elimination*, and *strength reduction*. While many of these are alleged to be architecture-independent, a closer look at embedded processor architectures shows that most of them are architecture-dependent. Following are two simple examples that illustrate the difficulties.

Figure 2.18 shows a very simple example of two statements which contain common subexpressions, b >> 2. If we consider a case where these statements are far apart in execution sequence, eliminating the common subexpression creates a local

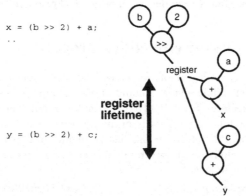

Figure 2.18 Common subexpression elimination increases variable lifetime

storage with a long lifetime. In the most likely case that the local storage is assigned to a register, this inhibits the use of the register for other purposes. In many embedded processors where registers are a scarce resource, the variable may need to be spilled to memory depending on how much local storage is between the two expressions. Another possibility in this example is to keep the variable b in a register and to recompute the value of b >> 2, which could possibly be a better choice if b is used further in the program.

The important point is that common subexpression elimination increases the *register pressure*, which may or may not improve the final code size or performance. An effective approach must take all the local storage requirements into account.

A second optimization example is constant propagation as shown in Figure 2.19. Figure 2.19 a) shows a sample C source with many statements containing a common constant expression: the variable x. A natural optimization is to propagate the constant 2 into the succeeding expressions to result in the C source shown in Figure 2.19 (b). However, we shall show that, depending on the architecture, even for this simple case this optimization leads to worse results.

Consider an architecture which supports three parallel operations: a data operation on a DCU (data calculation unit), a memory operation (load / store), and an

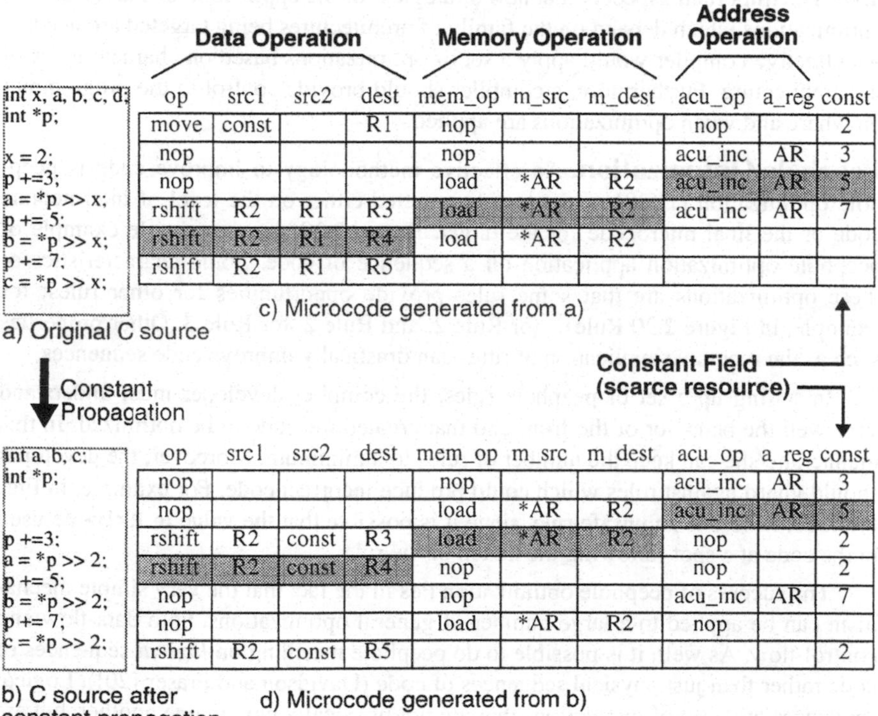

a) Original C source

```
int x, a, b, c, d;
int *p;

x = 2;
p +=3;
a = *p >> x;
p += 5;
b = *p >> x;
p += 7;
c = *p >> x;
```

c) Microcode generated from a)

	Data Operation			Memory Operation			Address Operation		
op	src1	src2	dest	mem_op	m_src	m_dest	acu_op	a_reg	const
move	const		R1	nop			nop		2
nop				nop			acu_inc	AR	3
nop				load	*AR	R2	acu_inc	AR	5
rshift	R2	R1	R3	load	*AR	R2	acu_inc	AR	7
rshift	R2	R1	R4	load	*AR	R2			
rshift	R2	R1	R5						

Constant Propagation

Constant Field ────
(scarce resource) ────

b) C source after constant propagation

```
int a, b, c;
int *p;

p +=3;
a = *p >> 2;
p += 5;
b = *p >> 2;
p += 7;
c = *p >> 2;
```

d) Microcode generated from b)

op	src1	src2	dest	mem_op	m_src	m_dest	acu_op	a_reg	const
nop				nop			acu_inc	AR	3
nop				load	*AR	R2	acu_inc	AR	5
rshift	R2	const	R3	load	*AR	R2	nop		2
rshift	R2	const	R4	nop			nop		2
nop				nop			acu_inc	AR	7
nop				load	*AR	R2	nop		
rshift	R2	const	R5						2

Figure 2.19 Constant propagation can occupy a vital resource: the instruction-word

address operation on an ACU (address calculation unit). In addition, it is common for an instruction-word to contain one constant field, as constants normally require at least 16 bits to be coded as an integer. Figure 2.19 c) shows a direct compilation of the source code a) into this type of architecture. Notice that this example results in an opportunity for a compact, pipelined execution of an ACU operation, load from memory, and DCU operation. For the same architecture, Figure 2.19 d) shows a direct compilation of the source code b). Notice that since the constant field is shared for all constant operations, the dependency on this resource causes breaks and stalls in the execution pipeline. Although the microcode of Figure 2.19 c) uses one more register (R1) to hold the constant, it is a much better solution than the microcode of Figure 2.19 d).

Optimizations on constants can have adverse effects on architectures with a scarce constant field resource. This fact has prompted the design of some architectures which provide long (e.g. 16 bits) and short (e.g. 8, 4 bits), as well as custom constant field formats in the instruction-word in an attempt to reduce the dependency on a long constant field, as well as free other bits for instruction coding.

Despite the fact that a multitude of standard optimizations prevail [1][30][9], their effects on microcode for real-time architectures can sometimes be counter-intuitive. The important aspect is that new strategies for the application of these standard optimizations which depend on the family of architectures being targeted are needed. An effective compiler would apply a set of optimizations based on characteristics of the architecture. Furthermore, a compiler should provide control to the programmer on where and when optimizations are applied.

Peephole Optimization. An effective methodology to improve code is peephole optimization [1][28], which can be applied either on the level of intermediate code or the final microcode for the target. Figure 2.20 shows a simple example of peephole optimization application on a sequence of code. Some characteristics of these optimizations are that some rules provide opportunities for other rules; for example, in Figure 2.20 Rule 1. for Rule 2. and Rule 2. for Rule 3. Other properties such as the recursive application of rules can drastically improve code sequences.

In setting up a set of peephole rules, the compiler developer must understand very well the behavior of the front-end that created the code to be optimized. In this manner, he/she can keep the number of rules to a minimum. Moreover, the developer should guard against rules which could produce incorrect code. For example, in Figure 2.20, Rule 2 is an unsafe rule, since it is possible that the value in R<b> be used in the code at a spot following the match of the rule.

The success of peephole optimization lies in the fact that the very simple mechanism can be applied to a large number of general optimizations, both data-flow and control-flow. As well, it is possible to do peephole matching on *logical* sequences of code rather than just physical sequences of code (Davidson and Fraser [20]). Logical sequences are a set of instructions that are not physically next to one another, but are connected through data or control-flow dependency. Since peephole optimizations are local, a well-structured matching mechanism can allow the application of hun-

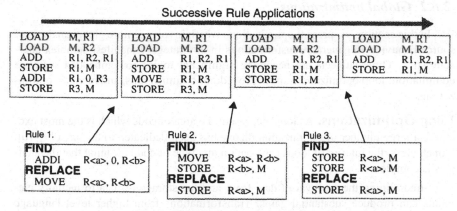

Figure 2.20 Peephole optimization example

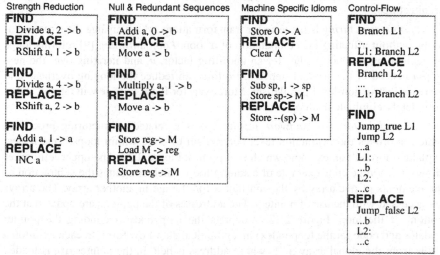

Figure 2.21 Some categories of peephole optimization rules

dreds to thousands of optimizations within reasonable run-time. A non-exhaustive list of categories of peephole optimization is shown in Figure 2.21.

The advantages of peephole optimization is the easy understanding and implementation. It can sometimes even improve compiler speed, since there is less code to assemble. The drawbacks of peephole and local optimizations is that they are machine-dependent and incomplete. Depending on the architecture, these optimizations are not enough to guarantee good code.

2.6.2 Global optimizations

In this section, we present optimizations which involve a more global analysis of source behavior. In general, transformations in this category manifest complex characteristics. Three subjects are becoming significant topics for embedded software compilation: loop optimizations, inter-procedural optimizations, and memory optimizations.

Loop Optimizations. As loops represent the area of code which is the most executed, a large number of optimization theory has been dedicated to this area [9][10]. Nonetheless, the interior of loops can contain any type of code meaning that the analysis is complex.

Streamlining the retrieval of data from and the storage of data to memory elements can produce substantial gains. Transformations from higher level language constructs like array and structure references to efficient machine-specific address generation are an important technology. This subject is treated in Chapter 5 for DSP architectures.

Loop restructuring is the term for transformations which change the structure of loops without affecting the computation of a loop. *Loop unrolling* [9] reforms loops by replicating the loop bodies for an unrolling factor, u, and iterating over the new step u instead of the original step 1. Unrolling can reduce the looping overhead and increase instruction-level parallelism. Moreover, for loops with few iterations, it can completely eliminate the loop structure.

Loop pipelining (or *software pipelining*) is a related restructuring procedure which improves the instruction-level parallelism of code within loops. This is best explained through an example which is shown in Figure 2.22. The upper left corner shows a C source code example of a simple loop which computes the subtraction of the elements of one array by the constant 3 for storage in another array. The arrays are accessed with the use of pointers. The addresses of the pointers are updated at the bottom of the loop. Figure 2.22 a) depicts the loop body (excluding the pointer updates and code for the loop index) in a graphical data-flow form. In each iteration a value from the global array a[], whose address is held in the pointer ap, is loaded into the register reg1, and subsequently the constant 3 is subtracted from reg1 to produce a value into reg2. This result is then stored to memory using the address in pointer bp. For a load-store architecture (RISC), this would mean that the body would cut into at least 3 instruction cycles (dashed lines): load-from-memory, addition, store-to-memory. For simplicity, we define each operation as taking 1 instruction cycle.

A partial pipeline is shown in Figure 2.22 b) where the load for the value pointed to by ap is executed once before the loop and in parallel with the subtraction operation. Also shown is an explicit representation in C code where the temporary values are to be placed in the reg1 variable which is declared to be of the register storage class. The pipelining has reduced the number of instruction cycles which are executed in the loop body. Although the microcode length has increased a little, the

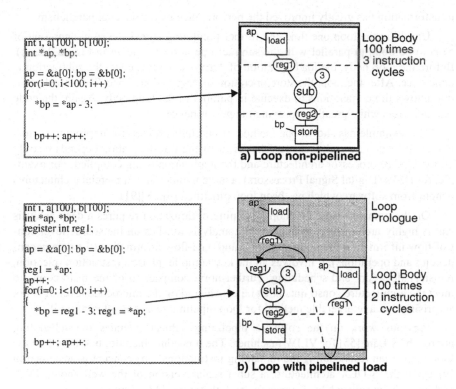

```
int i, a[100], b[100];
int *ap, *bp;

ap = &a[0]; bp = &b[0];
for(i=0; i<100; i++)
{
    *bp = *ap - 3;

    bp++; ap++;
}
```

a) Loop non-pipelined

Loop Body
100 times
3 instruction
cycles

```
int i, a[100], b[100];
int *ap, *bp;
register int reg1;

ap = &a[0]; bp = &b[0];

reg1 = *ap;
ap++;
for(i=0; i<100; i++)
{
    *bp = reg1 - 3; reg1 = *ap;

    bp++; ap++;
}
```

b) Loop with pipelined load

Loop
Prologue

Loop Body
100 times
2 instruction
cycles

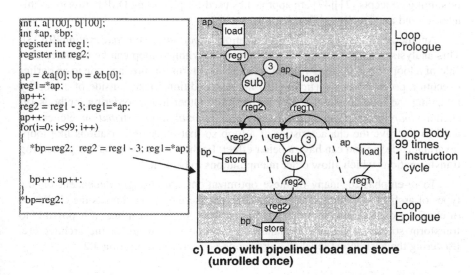

```
int i, a[100], b[100];
int *ap, *bp;
register int reg1;
register int reg2;

ap = &a[0]; bp = &b[0];
reg1=*ap;
ap++;
reg2 = reg1 - 3; reg1=*ap;
ap++;
for(i=0; i<99; i++)
{
    *bp=reg2;  reg2 = reg1 - 3; reg1=*ap;

    bp++; ap++;
}
*bp=reg2;
```

Loop
Prologue

Loop Body
99 times
1 instruction
cycle

Loop
Epilogue

**c) Loop with pipelined load and store
(unrolled once)**

Figure 2.22 Loop pipeling permitting arithmetic, stores, and loads in parallel.

transformation has greatly improved the performance by introducing parallelism.

Unrolling the loop one time and further pipelining allows the store operation of *bp to be done in parallel with the subtract operation as shown in Figure 2.22 c). Before the loop body, the load operation of *ap is done twice and the subtraction is done once. After the loop, the store operation of *bp is done once. This transformation allows three operations to execute in parallel in the loop body. Again, the code size increases while greatly improving the performance.

This example has shown one method of pipelining a loop for improving performance in load-store architectures. The effect of the transformation depends heavily on the type of processor architecture and the application being compiled. For example, for DSPs (Digital Signal Processors), a large number of commercial architectures benefit from software which has been loop pipelined [5][78][91].

On the compiler side, the loop pipelining optimization requires a deep analysis and is highly architecture dependent. The analysis touches on the source level control-flow information (the loop structure) and data-flow information (the data dependencies and operations). It involves all the key compiler phases: instruction selection, register allocation, and scheduling. Furthermore, compaction of the instructions to meet the architecture constraints must be considered on the microcode level. It is not surprising that a retargetable method for loop pipelining has not yet appeared!

Previous work on the software pipelining subject includes the scheduling approach by Lam [55] for VLIW machines. The procedure includes first unrolling the loop body, then rescheduling the remaining instructions. The concept is described in [9] using the S-DLX architecture, a super-scalar version of the well-known DLX architecture introduced by Hennessey and Patterson [43]. Loop folding [37] and pipelining concepts [21][47] are approaches used for pipelining DSP hardware architectures and are equally applicable to software.

Another simpler type of loop optimization is *loop-invariant code motion* [9][28]. This analysis determines whether a computation within a loop can be executed outside of a loop. *Code hoisting* [9][36] is the general term for moving code to an earlier execution point. *Loop unswitching* [9] moves conditional tests outside of a loop by repeating the loop structure for each condition. Other loop reordering methods [9] such as: *loop interchange, loop skewing, loop reversal, loop distribution,* etc. can be used to improve the characteristics of a loop so that other optimizations like loop-invariant code motion can have a better effect. Each of these are behavior-preserving transformations which allow other manipulations to be done.

To re-emphasize, many of these optimizations are strongly dependent on the types of architecture and strategies for applying these optimizations is the challenge of compiler construction for embedded processors. In the meantime, it is possible to transform source-level code by hand with a good knowledge of the architecture. Lowering the abstraction level of source code is discussed in Section 4.2.

Interprocedural Optimization. Modern high-level programming practices encourage modularity which suggests that small, well-bounded subprograms are better structured than large main programs. However, the mechanisms needed to support subprograms can often lead to inefficiencies. Two possibilities exist for the optimization of subprograms [28]:

1. In-line expansion of subprogram calls.

2. Optimization of called subroutines.

In-line procedures are subprograms whose code have been expanded to replace the call. Although similar to pre-processor macro expansions, they differ somewhat because of variable scoping rules. Languages such as C++ allow a programmer to suggest which subprograms are to be expanded in-line. Although ANSI C is somewhat more restrictive, it is possible to provide small extensions (#pragma) which serve the same purpose.

In-line expansion is predominantly a time vs. space trade-off, where the overhead of call-to and return-from subroutines are no longer needed. However, if called several times, the program code size can expand significantly. On the other hand, it is common to have subroutines which are called only one time which offers both a time and space improvement. Contrary to intuition, it is also possible to expand restricted versions of recursive subroutines. For example, a subroutine which computes the factorial of a constant could be expanded since the call depth is known at compile time.

If left to a compiler to choose which subroutines should be expanded in-line, a linking phase such as that used in C makes the procedure awkward. However, if all the functions are explicitly available, a compiler can make use of a subprogram call graph such as the example shown in Figure 2.23 to make the decision on which subroutines to expand. This graph simply denotes which subprograms call which at the execution of the program. Leaf (innermost) subroutines are the most likely candidates for expansion. Other useful measures for choosing candidates are calls from within loops, or better yet, execution frequencies generated from a profiler.

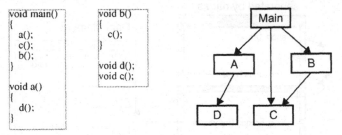

Figure 2.23 An example call graph.

When compiling the code within an in-line subroutine, the scoping information of variables is particularly useful. This marks the difference between in-line subroutines and preprocessor macros. While a compiler would need to determine the lifetime of variables into and out of macros, a well-written subroutine guarantees entry by parameters and exit by return values (excluding global variables).

Optimization of called subroutines is a very practical method in block-structured languages like Ada and Pascal. However, in languages like C, separable compilation and linking pose some problems. This is explored further in Section 4.1.3. When optimization is possible, in for example a restricted version of C, code saving techniques can be used. Most of the overhead in subroutine calls is in the *activation record* [1] and the local variables on the run-time stack. The activation record is a portion of code which is used to keep values such as the machine status (registers which are active), passed parameters, and return values. For subroutines with few parameters, it is possible to pass parameters in registers rather than on a run-time stack, should the architecture have enough registers. The same is true for return values; however, optimizing the assignment of registers can be a difficulty (discussed in Section 4.1.3). If a subroutine is determined to be non-recursive (by building a call-graph), a compiler can also make the trade-off of putting local variables in static memory rather than on the stack.

Memory Optimizations. Program memory of an embedded processor can be a particularly expensive part of the architecture, especially for single chip solutions. Efforts such as the narrowing of instruction words through encoding implies continual difficulties on compiler methods. An illustration of this is the limitation of absolute program memory addresses. Because of the short instruction words, an instruction-set which uses exclusively absolute memory addresses is limited in program size. Embedded processor designers have overcome this limitation in a number of ways. One approach is to provide near and far program calls and branches. A program memory can be organized in a set of *pages* as shown in the example of Figure 2.24.

While a paged program memory is a good solution for the hardware, it poses a number of challenges to the compiler developer. For code with good performance,

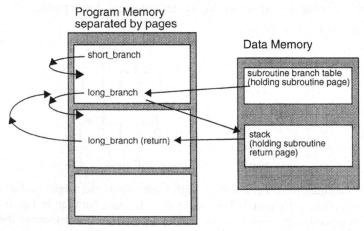

Figure 2.24 Paged program memory and an example subroutine management.

subroutines need to be allocated in memory in a fashion which reduces changing pages. Furthermore, long subroutines must be broken into smaller pieces so that each block fits into a page. Solutions to the problem are straight forward; however, an optimal solution is non-trivial. The example in Figure 2.24 shows a solution whereby subroutine page addresses are stored and managed by the compiler in a branch table. Each time one of these subroutines is called, the return page address is kept on the run-time stack.

The equivalent to program pages can also occur in data-memory, when a large amount of data memory is needed. For example, data *windows* can be used to organize the memory. Similar types of considerations must be taken in the compiler to minimize the time needed for the global storage and retrieval of data.

Another hardware solution to improve memory retrieval is the presence of *caches*. A cache is a temporary buffer which acts as an intermediary between program or data memory, improving the locality of the data. These are common to general-purpose computing architectures and appear on more sophisticated embedded processors. Approaches to improve the cache hit/miss ratio are beginning to appear for both program caches [101] and data caches [82]. Others have approached program memory reduction through code compression techniques [62]. The idea is to keep a program *dictionary* of frequently used sequences of code.

Allocation to multiple memories is a topic which arises for some DSP architectures, such as the Motorola 56000 series and the SGS-Thomson D950 (see Section 5.1). For these architectures, the parallelism is improved by allowing independent retrieval and storage operations on each memory. The implication for compilers is the need for memory allocation strategies which make best use of these resources based on the data-flow in the source program. This problem has been addressed in approaches which improve upon previously generated or hand-written assembly code by balancing the data in two memories [98]. However, techniques which are incorporated into the analysis phase of compilers are also needed. Still, practical considerations for memory allocation also need to be considered. These are discussed in Section 4.1.

2.7 Chapter summary

This chapter has presented a wide overview of modern compiler technology for embedded processors. It begins with a look at traditional software compilation and its relevance to the constraints of embedded processor architectures. A principal set of issues were identified as shortcomings with this approach including weaknesses in retargetability, ability to handle register constraints, capability over architecture specialization, and inherent compiler control for instruction-level parallelism and optimization.

Next, the most popular compiler with freely available sources was discussed: the GNU gcc compiler. As the compiler was designed for general purpose RISC architectures, some weaknesses appear for embedded processors including a dispersion of

information among various functions and macro definitions. The weakness of th
compiler for DSP architectures was further supported by the DSPStone and Berkele
Design Technology benchmarking activities of commercial compilers. The perfo:
mance of the commercial compilers based on gcc is described as less than acceptabl(

Next, the concept of compiler retargetability was discussed as a means of th
rapid set-up of a target compiler as well as an agency for architecture exploratioi
The promising avenue of architecture specification models and languages was exan
ined. Related work on specification languages includes: Mimola, a processor descrij
tion language on the structural level, and nML, a behavioral description language o
the instruction-level. Architectural models include the ISG of the Chess compile
and the mixed structural-behavioral model of the CodeSyn compiler.

Following, the three fundamental tasks of the compilation process were invest
gated: instruction matching/selection; register allocation/assignment; and schedulinj
compaction. Instruction matching/selection is predominantly done today by patter
matching and constructive methods. Register allocation/assignment is recognized ε
a critical compiler task and can be guided by classification schemes. Data-routin
approaches are a further assignment method for architectures with heavily coi
strained register resources and distributed register structures. Scheduling and coπ
paction approaches were then presented, illustrating the strong need for an effectiν
approach to exploit data movement and parallelism of a machine.

Finally a number of optimization techniques were described. The need for appl
cation strategies of standard optimizations was illustrated through discrepant even
in examples for very simple embedded processors. Peephole optimizations, loo
optimizations, interprocedural optimizations and memory optimizations were di:
cussed, all with regard to the challenges for today's embedded architectures.

Chapter 3:   Two Emerging Approaches: Model-based and Rule-driven

This chapter describes the function of two practical compilers for embedded processors developed and used by the author and colleagues in industrial contexts. While the two operate on very different principles, they each have their strengths with regard to the goals of retargetability and ability to generate efficient code. Naturally, each approach also has a certain number of weaknesses. However, each of the two tools has made particular contributions which have added to the understanding of the compilation problem for embedded processor targets. Furthermore, each approach has shown to have its place as an effective approach to compilation for embedded processors.

3.1 Overview of the concepts

Two recent approaches to compilation for embedded processors are shown in Figure 3.1. The first method, Figure 3.1a, uses a central processor model upon which all the phases of compilation are based. The source code is translated into an intermediate form, an explicit representation of the behavior of the source in a set of CDFGs (Control-Data Flow Graphs). The successive phases of compilation execute transformations upon this form to arrive at the final microcode. These transformations are done in a manner which satisfies the properties of the central architecture model.

The second method, Figure 3.1b, resembles quite closely the traditional approach to compilation as described in Section 2.1. The principal difference is the presence of processor information and retargeting rules. The rules provided by a developer allow the compiler phases to be reconfigured according to the architecture style. An open-programming concept provides an environment for the developer to build a target compiler.

In both approaches, information of the processor architecture drives the succession of the compilation steps. This feature is essential for embedded processors, where the architecture characteristics are unlike standard microprocessors. For example, processors for real-time reactive systems often contain a limited number of registers which are specialized for certain functions, as well as encoded instruction words, and special functions to communicate with the rest of the system. The compiler task of mapping onto these functions requires special attention.

47

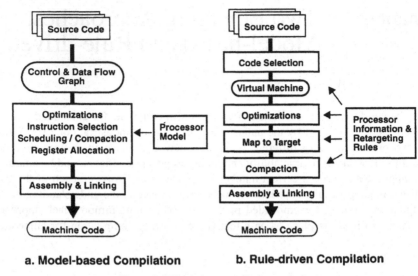

a. Model-based Compilation   **b. Rule-driven Compilation**

Figure 3.1 Emerging compilation techniques

3.2 The model-based CodeSyn compiler

The CodeSyn compiler was built mainly in response to a survey of the needs of designers of DSP systems for telecommunications. The details of the survey can be found in [83]. The survey, which was conducted among a number of design groups at Bell-Northern Research / Northern Telecom (Nortel), indicated several groups using both commercial DSP processors as well as application-specific instruction-set processors (ASIPs). Among other needs, the foremost was the requirement for compilers for both the commercial and in-house processors.

With respect to traditional compilation, the strengths of the CodeSyn system are in three main areas:

- A flexible instruction-set specification model which supports quick retargeting to new processors.

- An efficient pattern matching and selection approach which supports complex instruction recognition and utilization.

- Allocation of special purpose registers taking into account overlapping roles.

3.2.1 Overview of the approach

The overall flow of the process is depicted in Figure 3.2.

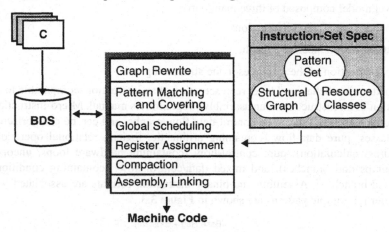

Figure 3.2 The CodeSyn compilation process

The compiler contains a set of modules including:

1. A source-level C parser. The C sources are converted into a hierarchy of Control-Data Flow Graphs (CDFGs) in an internal format called BDS (BNR Data-Structure).

2. A graph-rewrite module. This phase performs local translation of operations in the CDFG to operations found in the target architecture.

3. An instruction-set pattern matcher and selector. Pattern matching is formulated as the determination of all possible sets of instructions which can perform the function of the subject CDFG. The selection algorithm then chooses the best implementation from this set.

4. A scheduling module. This phase performs a coarse ordering of the patterns found in the matching and selection phase.

5. A register allocation and assignment module. Register classes are used in allocation, then local variables of the CDFG are assigned to specific physical registers of the architecture.

6. A back-end containing a compactor, assembler and linker.

Particular contributions have been made in the instruction-set matching/selection and the register allocation/assignment phases. These phases are detailed in Section 3.2.3 and Section 3.2.4 respectively.

3.2.2 Instruction-set specification

The CodeSyn instruction-set specification consists of a mixed behavioral and structural-level model composed of three main parts:

1. A pattern set of microinstructions.

2. A structural connectivity graph.

3. A classification of the resources in the structural graph.

A pattern is a behavioral level representation of an instruction, comparable to the description of an instruction in an assembly programmers manual. Micro-instructions are described as small pieces of control-data flow graphs, and can be categorized in three classes: pure data-flow (containing arithmetic, logical, relational operations, and address calculation); pure control-flow (containing hardware loops, unconditional jumps and branches); and mixed data/control-flow: (containing conditional jumps and branches). Assembly and binary instruction formats are associated with each pattern. Example patterns are shown in Figure 3.3.

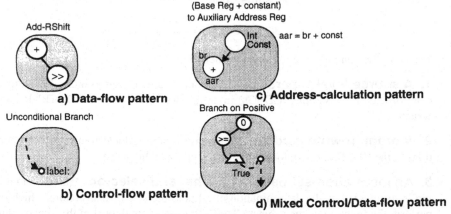

Figure 3.3 Example instruction-set patterns

An example of the structural graph and resource classification is shown in Figure 3.4. This structural graph is used by the compiler to determine the possible data movement through the processor. A register classification is used to categorize the registers in the architecture, which are typically categorized on two levels: a broad classification of general function and a small classification of specific function. Any number of overlapping registre classes are supported.

The three parts of the instruction-set specification are inter-related. The structural graph is built through a user specification of the relationship between register classes and functional units. Register class annotations are associated with the input and output terminals of each pattern indicating data-flow between classes. An example of this is shown in Figure 3.5. As well as allowing proper pattern matching, the register class annotation guides the register assignment algorithms to bind reads and

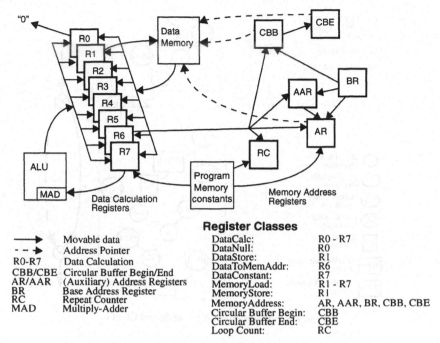

Register Classes

DataCalc:	R0 - R7
DataNull:	R0
DataStore:	R1
DataToMemAddr:	R6
DataConstant:	R7
MemoryLoad:	R1 - R7
MemoryStore:	R1
MemoryAddress:	AR, AAR, BR, CBB, CBE
Circular Buffer Begin:	CBB
Circular Buffer End:	CBE
Loop Count:	RC

Legend:

→   Movable data
- - ▶   Address Pointer
R0-R7   Data Calculation
CBB/CBE   Circular Buffer Begin/End
AR/AAR   (Auxiliary) Address Registers
BR   Base Address Register
RC   Repeat Counter
MAD   Multiply-Adder

Figure 3.4 Structural connectivity graph and register classes

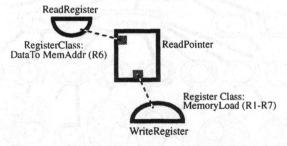

Figure 3.5 Register class annotation on the input/output of patterns

writes to physical registers in the architecture. One last relationship is the correspondence between operations in the pattern set with the functional unit which performs the operation in the structural graph.

3.2.3 Instruction-set matching and selection

An example source C code and corresponding CDFG are shown in Figure 3.6. The control-data flow graph (CDFG) is an explicit behavioral representation of the source C program containing a separable data-flow graph and control-flow graph. The left side of the CDFG (the data-flow graph) contains purely data-flow operations. Each of

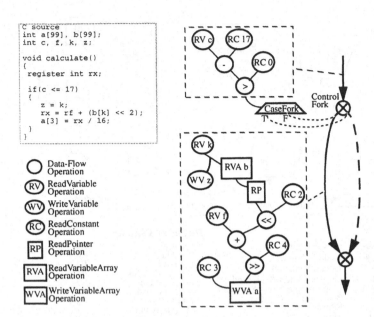

```
C source
int a[99], b[99];
int c, f, k, z;

void calculate()
{
 register int rx;

 if(c <= 17)
 {
    z = k;
    rx = rf + (b[k] << 2);
    a[3] = rx / 16;
 }
}
```

○ Data-Flow Operation

(RV) ReadVariable Operation

(WV) WriteVariable Operation

(RC) ReadConstant Operation

[RP] ReadPointer Operation

[RVA] ReadVariableArray Operation

[WVA] WriteVariableArray Operation

Figure 3.6 Example: source C code and CDFG

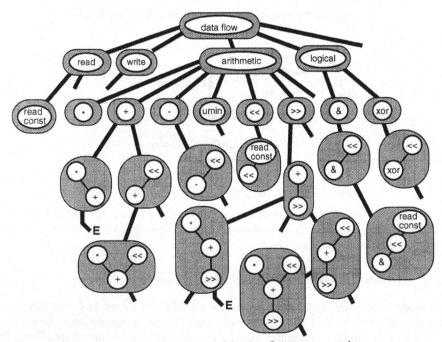

Figure 3.7 Root of section of the data-flow pattern pruning tree

these data-flow operations are bound to edges in the control-flow graph (the right side of the CDFG), indicating to which program flow the operations belong. At some points in the CDFG, data-flow operations pass information to the control-flow graph (`CaseFork`), indicating a control-flow decision point.

Notice that two graph rewrites have been done to the CDFG. The `if` condition has been rewritten to a subtract followed by a compare to zero, which allows matching to a `branch on greater-than` instruction; and, the `divide by 16` has been rewritten to a right shift by four. These are rule-based graph rewrites based on the specific capabilities of the architecture. Also, `register rx` is recognized as a local, temporary value and replaced by pure data-flow in the CDFG.

Data-Flow Patterns. Pattern matching is the task of determining isomorphic relationships between instruction-level CDFGs and the source CDFG. Instead of trying a match at every point in the CDFG with every pattern in the instruction-set, the number of attempted matches are kept to a minimum by a novel and efficient organization scheme, introduced in [65]. The pattern set is pre-sorted in a *prune tree*. This is a tree organized such that if a pattern does not form a match, then it is possible to prune the branch of the tree at this point for any further matches. It is guaranteed that no matches are possible beyond the prune point. Figure 3.7 shows the root portion of a data-flow prune tree for an example architecture. We have found in practice that it is possible to have enough organization of the prune tree so that the matching algorithm complexity approaches linearity with respect to the number of nodes in the subject graph.

At each node of the subject data-flow graph, the root of the prune tree is checked to see if it matches. If the root matches (it is a data-flow node), then the children of the root of the prune tree are checked. For each pattern that matches, the children are recursively checked for matches. For each pattern that does not match, the prune tree is pruned at this point for all the branches below this point. No further matches are possible below the prune point. At some points in the tree, it is also possible to slightly extend the tree into a dag such that the same characteristics of the pruning mechanism are retained. In Figure 3.7, the branches marked E may be reconverged into the respective pattern directly below, turning the tree into a dag. This incrementally improves the efficiency of the matching mechanism, since it increases the number of places where the branches may be pruned.

The approach is an efficient method to determine all the possible pattern implementations at all the points in the subject graph to allow an exploration of possible pattern selections. For example, Figure 3.8 a) shows a subject graph which has been matched to the patterns in Figure 3.7. Combinations of these patterns represent the possible implementations of the subject graph in a *covering* (i.e. instruction selection). Figure 3.8 b) and c) shown two possible coverings which can implement the subject graph. At first glance, Figure 3.8 b) would produce the better code as it is implemented in only two instructions; however, for the constant propagation problem of the type explained in Section 2.6.1 and illustrated in Figure 2.19, the covering of Figure 3.8 c) may be better depending on the context.

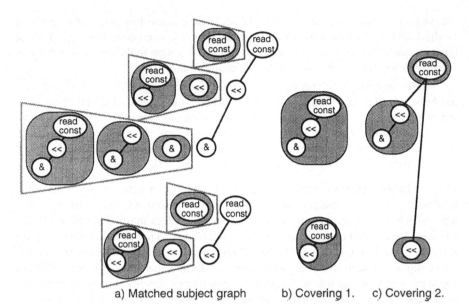

a) Matched subject graph   b) Covering 1.   c) Covering 2.

Figure 3.8 Subject CDFG with two possible coverings.

The important feature of the approach is that it provides pattern matches in a fashion that allows the matches to be propagated to other phases of the compiler to find the best coverings. The first covering algorithm developed in CodeSyn is dynamic programming as described in Section 2.3.1; however, the pattern matches are available for future improved algorithms in the phases of register allocation and scheduling.

Control-Flow Patterns. Figure 3.9 depicts an example control-flow pattern prune tree. The prune tree contains both patterns which are purely control-flow and others which have mixed control and data-flow. Although the prune tree is much shallower than its data-flow counterpart, the matching principles remain the same. Typically, the control-flow subject graph contains much fewer nodes and edges than the data-flow graph; therefore, matching time is manageable.

Notice that the conditional branch patterns shown here only include those which branch on true. It is possible to include the ones that branch on false, depending on the behavior of the C to CDFG front-end.

The matching algorithm for the control-flow patterns is analogous to the data-flow matching algorithm. In this case, matching begins by traversing all the edges of the control-flow graph, in contrast to the nodes in the data-flow graph. The matching algorithm determines isomorphic relationships of the control-flow graph with each control-flow pattern. If the pattern contains a mix of control-flow and data-flow, the algorithm directly calls functions used in the data-flow matching algorithm.

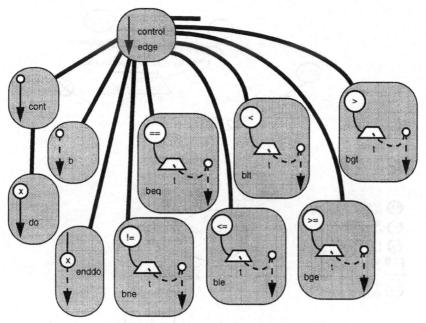

Figure 3.9 Control-flow pattern prune tree.

3.2.4 Register allocation and assignment

Benefiting from the annotation of register classes on the input and output terminals of patterns, the allocation of registers is done by calculating overlaps in the classes. Following the data-flow between nodes, candidate register sets are calculated from the intersection in each annotated register class.

Shown in Figure 3.10 is a matched and covered CDFG in preparation for register allocation. Instructions have been identified which perform the implementation of the operations in the data-flow graph. In addition, WriteRegister and ReadRegister operations have been explicitly placed between the operations, with the annotation of register classes found on the input and output terminals of the patterns. Following, candidate register sets are calculated from the intersection of register classes between WriteRegister operations and ReadRegister operations as shown in Figure 3.11.

The task of register assignment is to determine specific physical registers as the input and output operands of the instructions. Intersecting register classes containing no candidate registers (Figure 3.11 c) can only be resolved by moves to available registers and spills to memory.

The approach to register assignment is greedy in nature and geared toward giving priority to registers dedicated to specific tasks. The general procedure is as fol-

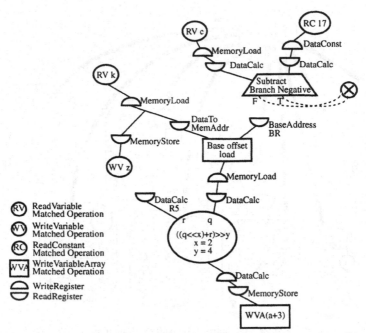

Figure 3.10 Matched and covered CDFG - preparation for register allocation.

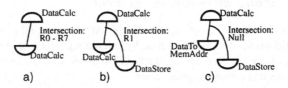

Figure 3.11 Register allocation - calculation of candidate register sets

lows and the details are described in [66]:

1. Assignment begins at intersection points which have register classes containing only one member register.

2. Assignment for the intersection points in Step 1. are ordered based upon lifetime and number of reads. Assignment begins with the shortest lifetime and fewest number of reads.

3. Steps 1 and 2 can lead to register assignment conflicts which are handled by greedily inserting register moves.

4. The remaining candidate intersections are assigned by a left-edge algorithm (see Section 2.4.2) enhanced for overlapping register classes. The procedure begins with the most constrained class (fewest members) and continues to the least constrained class (most members).

At points where there are no available registers, spills to memory are inserted. This is nontrivial since the DataStore class is needed and it may be a dedicated register (R1) in some architectures (see Figure 3.4). This problem has been resolved by placing priority assignment orders on the registers in each class. Registers which are members of the DataStore class are the last on the priority list to be assigned.

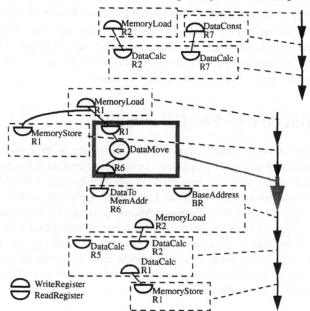

Figure 3.12 Scheduled CDFG with registers assigned

Figure 3.12 shows the same example as Figure 3.10 focusing on the ReadRegister and WriteRegister operations. The operations in the CDFG has been coarsely scheduled using list methods. As well, register assignment has been completed. Notice that a conflict move has been inserted to resolve the two ReadRegisters which have been assigned to special purpose registers (R1 and R6), and are written by a common WriteRegister. Apart from compaction, at this point the CDFG is completely mapped to the instruction-set of the architecture. Assembly code format instructions which have been associated with the patterns are emitted along with the assigned registers. This results in sequential assembly code. Compaction is done in a separate phase as is explained in Section 3.3.5.

3.2.5 Assessment of the approach

A model-based approach to retargetable compilation has a number of advantages. The instruction-set model provides a central core whereby all the phases of compilation can rely on architecture information. This naturally augments the retargetability of each algorithm. In particular, the CodeSyn compiler has shown that it is possible to use enhanced algorithms which can perform efficient pattern matching and selection in addition to register allocation and assignment for special-purpose registers. An intermediate representation which is rich with control and data-flow information is useful for optimizing the mapping onto the instruction-set.

On the other hand, a full control and data-flow representation can also be quite a heavy set of information. It contributes to the needs of maintenance and could also be a barrier to compilation speed, should the source programs become lengthy.

An efficient mapping to a target architecture is possible with a model-based approach; however, it necessitates that the architecture lie within the boundaries of that model. Any architecture peculiarities of a new target which are not anticipated must be handled by improving the retargeting algorithms.

3.3   The rule-driven FlexCC compiler

In the case of providing a compiler service, the desire may be to provide compilers for the widest possible variety of processors. In the absence of an automatically retar-getable compiler system which is applicable to any architecture, a flexible *compiler development environment* has benefits. One such approach is the rule-driven approach, currently in use at SGS-Thomson Microelectronics for in-house embedded processors. The phases of this compiler have been restructured in such a manner as to make it easier to reprogram the compiler for new targets. Building upon traditional compiler techniques, this programming environment for compiler development can improve the time of the retargeting process, as well as provide a manner to reuse compiler strategies and experience.

3.3.1 Overview of the approach

This approach to compilation was first presented by Gurd in [38]. It is based on step-wise progressive refinement whereby each phase of compilation is formed upon an open programming concept. This programming environment allows a compiler development team to build rapid prototypes using a well-defined train of tools. The compilation process is roughly based upon the traditional view of a compiler as explained in Section 2.1 and shown in Figure 3.13.

A compilation is divided into four main phases, which are shown by example in Figure 3.14 and summarized as follows:

1. Virtual code selection. The developer defines a virtual machine which resembles in functionality the instruction-set of the real machine, but is sequential in operation. Those processors with parallel execution streams would be simplified to one stream. The virtual machine description contains two main parts:

1.   a description of resources including register sets and addressing modes.

2.   a set of code selection rules.

2. Optimizations. Instructions for the virtual machine may be passed to a series of optimization routines, such as a peephole optimizer, whose principles are

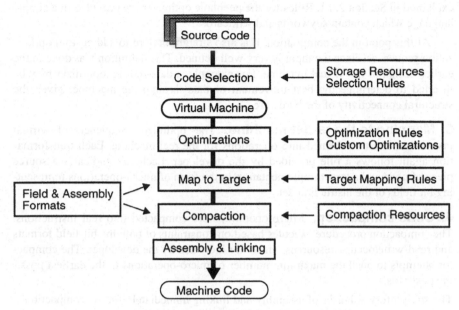

Figure 3.13 Overall flow of rule-driven compilation

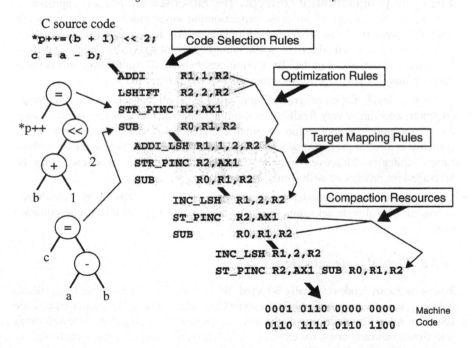

Figure 3.14 Refinement steps in a rule-driven compilation.

explained in Section 2.6.1. Rules for the peephole optimizer are provided in a simple language which contain keywords and wildcards.

At this point in the compilation, it is also straightforward to add custom optimization sequences, since the input is very well defined. This definition was done in the earlier design of the virtual machine. For example, a data-routing optimizer may be inserted to determine the best movement of data through the machine, given the structural connectivity of the hardware.

3. Mapping to the target machine. The optimized sequence of virtual instructions are transformed into operations for the real machine. Each transformation again follows a rule provided by the developer. Each rule indicates a source piece of code and a target implementation in the form of micro-operations representing bit fields of the instruction-set.

4. Code compaction. Micro-operations are compacted into real instructions. The compaction procedure executes based on constraints of both the bit-field formats and read/write/occupy resources which are indicated by the developer. The compactor attempts to push the maximum number of micro-operations to the earliest possible positions.

The straightforward tasks of assembly and linking immediately follow compaction.

The open programming concept. The rule-driven compilation approach is built upon the concept of an open programming environment. All the rules are defined in well-structured programming languages. At each point in the compilation, primitives are provided which allow the identification of cases which may appear in the source. These cases can then be manipulated by a set of control-flow functions in the programming language for the emission of code for the next step.

A high level of open programming is provided at all the steps of the rule-driven compiler, allowing a very flexible development system. The user is able to retarget the system upon the application of suitable mapping functions. The quality of the compiler is directly proportional to the amount of development time spent on optimization strategies. Moreover, previous compiler development experience may be leveraged for processors with similar features.

In the following sections, we detail three of the key steps in the rule-driven approach: virtual code selection, target machine mapping, and microcode compaction.

3.3.2 Virtual code selection

To re-emphasize, code is initially selected for a virtual machine based on two criteria: a description of resources including registers and addressing modes, and a set of code selection rules. The definition of the available register sets classifies these resources into functional categories, for example, it indicates which C data-types each register may hold. The definition of the addressing modes indicates the manner in which variables are to be retrieved from memory. The addressing modes may be defined using a

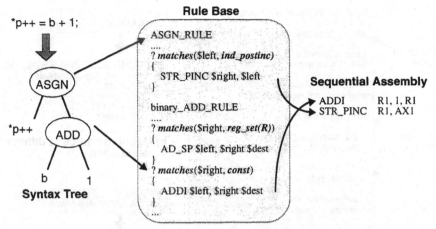

Figure 3.15 Sequential code selection

combination of data-type sizes and constant offsets to describe modes such as immediate, direct, register indirect, etc. Thus, both RISC and CISC machines can be supported.

For the code selection rules, the developer defines the mapping between the C code onto the virtual machine instruction set. The compiler developer has at his/her disposal a programming language which contains a set of high-level primitives corresponding to information which is generated as syntax trees of the source program. For each operation which may occur in a syntax tree, the developer provides a rule for the emission of code for the virtual machine. This rule will be triggered upon matches to the source code and executed at compile time. An example is shown in Figure 3.15. Syntax trees are constructed from an analysis of the source program, and each tree triggers a rule depending upon the operations of the nodes. The developer can then provide case functions on what code to emit depending on the properties of the tree. The programming language contains a large number of features such as the ability to call rules from other rules and recursively call rules.

This approach allows the developer to provide simple rules for the majority of cases and more complex mappings for special features of the architecture. For example, the developer may restrict the use of certain registers whose function are constrained by the architecture. This is important to support the special-purpose registers found in embedded processors.

Register assignment within register sets is performed after code selection using a coloring approach. The approach uses standard techniques as described in Section 2.4.2, in a manner which satisfies the constraints imposed by the code selection rules.

3.3.3 Target machine mapping

This step performs a refinement of the sequential operations for the virtual machine

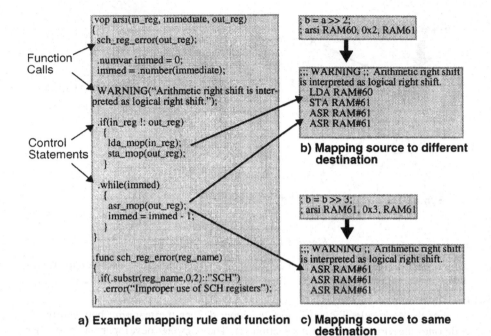

Function
Calls

Control
Statements

b) Mapping source to different destination

a) Example mapping rule and function   **c) Mapping source to same destination**

Figure 3.16 Target mapping example

into instructions for the real machine. In many cases, this can be a simple one-to-one mapping; however, a C-like language offers the capability to manipulate this mapping based on values generated by the previous steps. The most common values used for manipulation are the parameters of the assembly code. For example, for the assembly instruction, INC R1, manipulations may be made depending on the parameter *R1*. An example which illustrates some of the features is shown in Figure 3.16. The example shows a rule driven from an arithmetic right shift operation. Several features such as function calls, control statements, and local variables provide a rich palette whereby the programmer can manipulate the generation of micro-operations. The example of Figure 3.16 b) shows a mapping which fires the contents of the if(in_reg !: out_reg) statement in the mapping rule, while the example Figure 3.16 c) does not activate the statement. The example also shows the possibility of a compile-time expansion of constant values to multiple instructions (while loop).

3.3.4 Microcode compaction

The compaction phase of the rule-driven compiler is based upon well-known methods as described in Section 2.5. Some features that make it a practical approach is the ability to retarget the process to varying styles of architectures. The compactor attempts to pack micro-operations as tightly as possible within the given constraints.

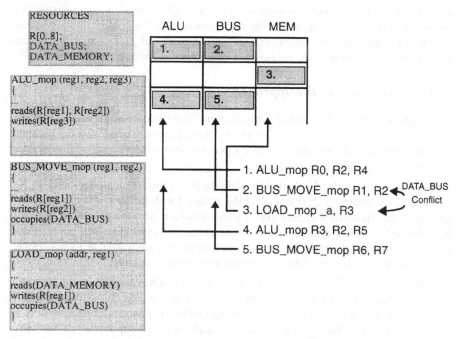

Figure 3.17 Programmable micro-operation compaction

Those constraints come in two forms: the bit-fields in the micro-instruction-word which are assigned by micro-operations; and a set of resources defined by the programmer. These resources usually include architectural storage units such as the registers and memories, as well as shared transitional units such as busses. In the definition of micro-operations, the programmer is obliged to indicate which resources are written-to, read-from, and occupied. Thus, the compaction algorithm is able to obey all the data-flow dependencies, as well as the resource restrictions of the machine.

An example is shown in Figure 3.17, where three types of micro-operations (mops) are declared, each defined to read, write, and occupy certain resources. As the compactor proceeds through the list of micro-operations 1 to 5, it pushes each mop into its respective field. If there is no resource conflict, a mop can be placed in parallel with (e.g. Figure 3.17: mop 1 and 2) or even past a previously placed mop. Resources include registers and memories which keep data-dependencies in order, but also include architecture constraints like the occupation of functional units or busses (e.g. Figure 3.17: mop 2 and 3). Furthermore, non-existent resources may be defined to specify unusual architecture constraints.

While giving a flexible view to the programmer on how the compaction procedure is executed, the onus is on the developer to validate that all the resources have been correctly declared and used in each micro-operation. However, granted that the approach is flexible as a compaction procedure especially for VLIW machines, it does not provide a straight-forward solution to highly encoded micro-instruction words since the bit-fields represent the first constraint of the compaction algorithm. To address highly encoded instruction-sets (see Section 1.3.2), a compaction procedure which takes into account the *meshing* of instruction formats must be used.

3.3.5 Assessment of the approach

Although the rules for this type of.compiler must be written by an experienced developer, the retargeting time is relatively short. Experience has shown that the retargeting time typically falls between one to six person-months. More on the effort needed for retargeting is discussed in Chapter 6.

The main strength of rule-driven compilation is the inherent flexibility of the approach. The compiler developer has the means for describing specific rules and strategies for efficiently mapping higher level constructs onto the processor, based on his knowledge of the architecture idiosyncracies. Standard rules are put in place based on previous compiler experience, and primitives are available to manipulate the compiler for new architectures with unforeseen specialization.

When compared to a traditional compiler approach, the rule-driven approach allows for faster development time through the ability to retarget at each phase of compilation. The quality of the results depend on the compiler development effort. In the few cases when the code quality is inadequate, a custom optimization module is incorporated with little effort.

When compared to model-based retargetable compilation approaches, the rule-driven approach requires retargeting development time; whereas, in principle, a model-based compiler requires only a small change to the model to arrive at a new compiler. However, in our experience, the retargeting time is compensated by the applicability of the rule-driven approach to a very broad set of processor architectures, from low-end microcontrollers to VLIW DSPs (Very Large Instruction Word Digital Signal Processors). In addition, architecture specific idiosyncracies may be handled by case-by-case development strategies.

3.4 Chapter summary

This chapter has presented two emerging examples of compiler systems aimed at application specific instruction-set processors (ASIPs). What sets them apart from traditional approaches is an instruction-set specification containing information about the processor. In the case of the model-based approach, the specification is a set of architectural properties which include the functional units, connectivity of storage resources, and a set of instructions. In the case of the rule-driven approach, the specification is the combination of processor information including instruction formats plus a set of transformation rules used at each phase of the compilation.

Both approaches have a set of strengths which make each a compelling approach for today's embedded architectures. A model-based approach can potentially provide a highly optimized mapping based entirely on properties of the architecture. The model would also allow the designer to explore variations on the architecture by making small changes to the specification. The rule-driven approach can potentially provide a very wide retargeting range for a service-based compiler group. The flexibility of rules allows the developer to explore new compilation strategies for each new target processor.

Chapter 4: Practical Issues in Compiler Design for Embedded Processors

This chapter discusses pragmatic issues in setting up a compiler environment for embedded systems. While the techniques presented in Chapter 2 form the basis of a compiler system such as the examples in Chapter 3, many other factors must be brought into consideration for a usable development environment. These factors may include:

- Language support: What ingredients of a programming language should be provided to the user?

- Embedded architecture constraints: What facilities should be provided to the user to control the specialized architecture?

- Coding style: What abstraction of coding style should be supported? What are the trade-offs?

- Validation: What level of confidence will be provided with a retargeted compiler?

- Source-level debugging: How does debugging on the host fit in with debugging on the target?

These as well as other practical considerations are often more important to the design engineer than simply the base technology of the design tools. Only a complete development system allows for efficient embedded software design.

4.1 Language support: choosing the right subset and extensions

In the embedded industry today, the language of choice is C [33]. While there exist languages more suitable for certain domains (e.g. Silage/DFL [44] in the DSP domain), C remains the most widely used high-level language in embedded processors mainly because of the wide availability of compilers and tools on workstations and PCs (linkers, librarians, debuggers, profilers, etc.). Furthermore, many standards organizations such as ISO (International Standards Organization) and the ITU (International Telecommunications Union) provide executable models in C. Some examples of these are: GSM (European cellular standard), Dolby (audio processing), MPEG (video and audio processing), H.261/ H.263 (videotelephony), and JPEG (still picture processing).

While C is an expressive language, many limitations [105] are imposed for the use in embedded applications:

- limited word-length support. The fixed point support in C is limited to 8 bit (char), 16 bit (short int), and 32 bits (long int). While this is sufficient in applications such as speech processing, it is insufficient in many other embedded applications, for example in audio processing where typically 24 bit types are needed and image processing where much larger data-types are needed.

- a limited set of storage classes. In many DSP systems, in addition to multiple register files, there are at least two data memories as well as a program memory. For certain applications, there can be even more [64]. ANSI C provides only the auto, static, extern, and register storage classes [33], which are insufficient in providing the user control over where the data is to be placed.

- a fixed set of operators. Embedded systems may have hardware operators which do not correspond directly to the operations found in C.

- limited scheduling and parallelism. The semantics of C impose a fixed schedule on the order of operations, which can be limiting for wide instruction machines (e.g. VLIW). Although it is possible to take a liberal interpretation of the schedule maintaining correct functionality, this is often difficult in the presence of pointers and the aliasing problem [1].

- separate compilation and linking. C allows modules to be compiled separately; the modules are then linked together in a separate phase. In the presence of limited register resources, this imposes an obstacle to efficient inter-procedural optimization, such as the passing of arguments in registers.

While there are limitations in C, the practical solution is to work within the constraints to provide the right compiler support for the architecture at hand. In many cases, the above-mentioned constraints are not limiting, while in others the difficulties can be managed. The way to work within the limitations is to make good choices about the levels of support. For example, a subset of C could be chosen to allow a certain optimization. Once the user steps outside that subset, this optimization is no longer guaranteed. In other cases, a minimal extension to the C language gives the features desired. If handled carefully, this does not destroy the compilability of the code with other C compilers.

4.1.1 Data-type support

Whereas general computing processors have evolved to support the data-types found in a high-level language, the architectures of embedded processors usually support only the data-types needed for a set of applications. These are typically few in number. Consequently, in the design of a retargeted compiler, all the tasks are made simpler if the number of supported data-types are kept to a minimum, since they must finally be mapped onto the data-types supported by the architecture. While it is possible to support larger data-types by providing libraries of larger data-types built upon smaller data-types, this elevates all the tasks in developing, maintaining, and validating the firmware development environment. Furthermore, the embedded processor programmer is most concerned with performance. Working at a level where the data-types match the register and memory widths of the architecture is the most natural level at which the designer can guarantee real-time performance. While working at a higher level (e.g. larger data-types) may simplify the programming, it is a secondary concern of the designer after the guarantee of meeting real-time performance constraints.

While there is a limited number of fixed data-type support in C, it is not often the case where the architecture supports an extensive number of fixed data-types. It is therefore possible to *re-map* certain C data-types to the types needed for the application. For example, in audio applications, a word-length of 24 bits is commonly used for sampled data. Whereas the 24-bit data-type does not exist in C, the 32-bit data-type may be used instead. This is assuming that the 32-bit type is not needed as well. The data-type may be reinterpreted by the retargetable compiler.

This approach leaves no visible consequence on the side of the target compilation path. However, within a methodology where the host compilation path is kept in sync with the target compilation path, the equivalence is broken. Section 4.3 describes the importance of keeping the two paths equivalent. With this approach of reinterpreting certain data-types, the operations of the host compilation would not match the operations of the target compilation. In our experience, this can be solved in one of two ways:

1. the use of built-in functions (Section 4.2.1) and the provision of a bit-true library (Section 4.3.2). Built-in functions provide a common interface for compilation in both paths, allowing the correspondence between host compilation and target compilation to remain intact.

C operator	Built-in function
a = b * c;	a = MULT_24(b, c);

Figure 4.1 Using built-in functions for unsupported data-types.

The function definition of the built-in functions may be provided in a bit-true library for compilation on the host.
While this approach may appear cumbersome for specifying operations, it is manageable, especially if the number of data-types which differ from those in ANSI-C are few.

2. extending the host compilation to the use of C++ data-types and operator overloading. The operators of C++ can be overloaded to provide the bit-true operations depending on the data-type. While, the retargetable compiler still treats the

source as C, the mapping of the data-types is interpreted according to context as shown in Figure 4.2. The equivalent compilation on the host is interpreted using C++. Ideally, these C++ data-types could also be used by the retargetable compiler to offer a extendable set of data-types for any application. This approach is proposed in the Chess compiler developed by IMEC / Target Compiler Technologies [34].

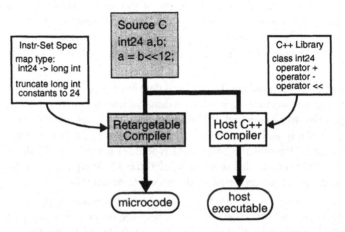

Figure 4.2 Supporting C++ data-types with a retargetable C compiler.

4.1.2 Memory support

An increasingly common method to improve memory support in C, is to extend the storage classes in ANSI C to those needed for the architecture. For example, if two RAM data memories exist on the architecture, a storage class specifier _MEMORY1_ could indicate the first memory and _MEMORY2_ could indicate the second memory. This allows the designer to choose the location of his data variables. This is a pragmatic solution in contrast to providing a memory allocation algorithm in the compiler. In our experience, if more than one data memory is available on the architecture, this was done explicitly by the designer to organize the architecture for a specific reason. For example, this may be to separate different types, memory-mapped I/O, or ROM from RAM. In each of these cases, the designer has a preconceived idea of where the data is to be placed. However, if the separation of memories is purely for reasons of speed and there are no large differences between two memories, the compiler would benefit from a memory allocation algorithm (see Section 2.6.2).

In the case of separate memories of RAM and ROM, it is also possible to use the type qualifier const to identify ROM values, instead of specifying the storage class. The compiler simply needs to intercept these variables for placement in ROM.

Extending the methodology further, the compiler can treat memory-mapped input/output (I/O) in the same manner. Memory-mapped I/O are locations in memory which are actually register interfaces connected to the core's hardware peripherals. Their addresses in memory form a convenient identification system for the processor. Storage class specifiers can also be used to identify memory-mapped I/O. However, in this case there are two further important characteristics of a memory-mapped I/O variable. First, as a specific address is required, the user must be able to force the variable into that location in memory, either through a #pragma or with an ANSI C extension. Secondly, the compiler must be made aware that it cannot remove accesses to the variable through optimization. The type qualifier volatile accurately defines this characteristic.

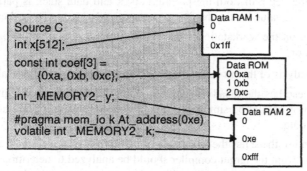

Figure 4.3 Storage class specification and type qualification for multiple memories.

For memory-mapped I/O, sometimes further consideration about the behavior of the target compiler must be taken into account. The functionality of the architecture may place constraints on the compiler. For example, depending on the protocol used for memory-mapped I/O accesses, scheduling and compaction of operations with the memory-mapped I/O access may be forbidden or in other cases obligatory.

4.1.3 Procedure calls

For small embedded applications, where the implementation of a stack is not justi-fied, the extended use of pre-processor macros can give the user the appearance of procedure and function calls. Of course, the designer then pays the penalty of the in-line code expansion for each macro use. Alternatively, it is possible to provide one level of function call with minimal hardware support. This requires simply a jump-to-subroutine/return instruction pair and one register to guard the return program address. In our experience, we have found that for minimal architectures, this is a useful hardware addition despite the limitations of parameter passing only in regis-ters and simply one level of calling.

When one level of procedure call is insufficient, certain hardware considerations need to be taken into account. A program stack provides the necessary functionality for nested subroutine calls. For parameter passing and local automatic variables, a data stack needs to be provided. The program stack and data stack may be merged into one if the program and the minimum addressing unit of the data are of the same bit-width. If they are not of the same width, using the same stack could be a consider-able waste in memory.

If there is more than one data memory with different bit-widths, placement of the data stack is a difficult issue. If placed in the largest memory, it supports all the data-types, but at a waste when using the smaller data types. If placed in a smaller mem-ory, then the compiler must be able to store and retrieve the larger data-types without losing precision. Providing more than one data stack is another solution; however, very few of today's compilers are able to handle more than one data stack.

Usually, the depth of both the program stack and data stack is parametrizable. Choosing the best value for the depth can pragmatically be done by either:

1. simulation on the workstation while tracing the number of calls and automatic variables, or

2. a static analysis of the procedural call tree and number of local variables.

The first of these has the disadvantage that it may take a long time, given that the entire application should be simulated. In addition, the result is guaranteed only for the given test data.

The second of these has the advantage of being fast; however, to be accurate, the code generated from the target compiler should be analyzed to determine the number of automatic variables used locally. This value can vary greatly depending on the number of and the constraints on the available registers. In either case, the possibility of recursive procedure calls may cause the stack depth estimates to be inaccurate. However, many compilers for embedded processors disallow recursive procedures.

As mentioned at the beginning of this section, separable compilation imposes restrictions on the results of the target compiler. As the compiler must account for procedures being compiled in any sequence, less optimizations are possible. For example, an interprocedural optimization such as the passing of parameters in regis-ters are obliged to follow a convention (e.g. first argument in register 1, second argu-ment in register 2, etc.). This is to allow any future and prior calls to know the location of parameters. Consequently, the architecture constraints on the use of regis-ters may mean that these parameters must move location before being used. The example in Figure 4.4 illustrates this restriction.

Consider the procedure filter shown on the top left corner of Figure 4.4. It contains a call to the procedure decode, which is not yet compiled. This is permit-ted in C since the prototype for decode appears prior to the call. If the compiler chooses to pass the parameters of the call, x and y, by registers to the procedure decode, it must use a convention such as x into R1 and y into R2, since it has no previous knowledge of the interior of the function decode.

Now consider the procedure decode, which is compiled after filter. The parameters a and b are passed by convention, a into R1 and b into R2. However, in the context of embedded processors, constraints on the use of registers can appear through architecture specialization. This allows the designer to streamline the instruction-set and hardware. The data-path shown on the right of Figure 4.4 shows

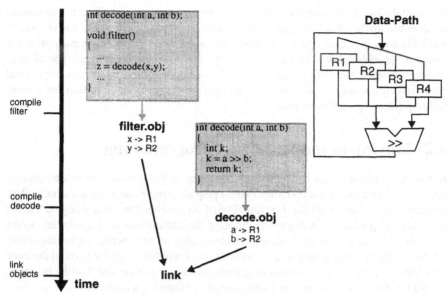

Figure 4.4 The implications of separable compilation on interprocedural optimization

an example of this type of restriction. The shift operation can be performed on any register R1, R2, R3, or R4; however, the number of shifts must reside in register R4. Therefore, this implies that in our example, the parameter b must be moved from register R2 to register R4 prior to the shift operation k = a>>b. This is an inefficiency inherent in the support of separable compilation and cannot be avoided. In some cases, the penalty can be great, for example if the call were located in the body of a critical loop.

This inefficiency can be side-tracked if the user is willing to use a lower coding style, as will be described in Section 4.2.2. For example, the user could assign the location of variables to registers himself, knowing how he intends to use them. This implies much more thinking on the part of the programmer and much less portable code.

Another implication of separable compilation is in memory assignment: the automatic positioning of variables in memory based on their use. For example, the offset assignment algorithms in [60] and [62] propose the positioning of variables in memory based on their use in a data-flow context which allows a minimum of addressing arithmetic. If these algorithms were to be used for global variables, this precludes the support of separable compilation. This is because they assign memory positions based on the use of variables within procedures. For procedures which are compiled after this memory assignment, the variables will not be in optimal positions. The algorithms do apply to local static variables; however, in C, local variables usually reside in registers or are put on the run-time stack to support recursion.

In order to allow optimizations like the previous two examples, it would be necessary to disallow the separable compilation of C. This would have a large impact on the compilation times, requiring the equivalent of complete compilation and linking of an entire application every time. In addition, it would disallow the use of object-level libraries which is often a methodology used when providing hand optimized assembly-level functions. However, this is the only way to allow these type of optimization possibilities, which may be a useful methodology should the amount of total embedded firmware remain low. Also it could be provided as an option for a final, fully-optimized compilation pass.

4.2 Moving beyond assembly programming

In this early phase of the acceptance of compiler technology for embedded processors, it is imperative that a compiler system provide simple mechanisms which allow programmers to reach all the functionality of an architecture. If a designer cannot meet his/her performance objectives through the capabilities of a compiler, he/she must have the ability to reach the equivalent quality of hand-written assembly code, while maintaining the ability to use a high-level of programming for parts of the code which do meet the objectives. For pragmatic reasons, it is essential that the bridge to higher levels of automation always be crossed as smoothly as possible.

4.2.1 Built-in functions

A built-in function is a compiler-recognized function which is mapped directly onto a set of instructions of the processor. These allow the execution of operations which are not found in C, for example: interrupt instructions, hardware do-loops, wait mechanisms, hardware operators, co-processor directives, the setting of addressing modes (modulo, bit-reverse, etc.). In addition, built-in functions are useful for providing access to specialized hardware to which compilation is done inefficiently. In providing built-in functions, the compiler developer must ensure that any optimizations or manipulations of the control-flow and data-flow operations do not interfere with the behavior of the function.

4.2.2 Coding styles on different levels

A retargetable compiler should allow coding on various levels of abstraction as well as the mixing of these levels. This is to allow the designer to reach all the functionality of his/her processor, at perhaps the expense of code portability, when the compiler cannot provide it. Our experience has shown that while the development effort on optimizations is important to achieve a more portable level of code [69], the effort in ensuring that the compiler can handle lower coding levels is essential.

For C, we define four levels of coding styles as follows:

1. High Level
    Behavioral ANSI C: This level is characterized by the use of variable, array, structure references and all the operations available in C.

2. Mid Level

   This level allows the use of built-in functions. Any arrays or structures that are declared in memory must be accessed by pointers. Variables and pointers may be allocated into extended storage classes and register sets.

3. Low Level

   This level allows the user-assignment of variables and pointers to specific registers.

4. Assembly Level

   This level allows the programmer to write in-line assembly code directly in C code.

Level 1 is the goal for compiler technology. This is the level at which all optimization and retargeting capabilities should aim as it provides the most abstract and portable source descriptions. It is also the level at which a programmer can freely write algorithms without being concerned with the underlying hardware.

Level 2 is reached mostly by capabilities within a compiler. Built-in functions are exactly like any other C functions, but are mapped to specific instructions by the compiler. Extensions to the storage classes of C allow allocation into memories (Section 4.1.2) and register sets. The allocation to register sets allows the compiler to perform better register assignment.

Level 3 is arrived upon by small syntax extensions to ANSI C. These *C-like* extensions declare variables to reside in specific registers. It is important to note that C code which remains within these first three levels allows compilation on the host given the provision of the proper masking of extensions and a bit-true library.

Level 4, the assembly level allows assembly code to be mixed with C code. A function-like interface allows the programmer to directly write assembly instructions for the processor within the framework of the C code.

Examples for these 4 levels are shown in Figure 4.5. The high level example shows the use of behavioral C constructs like the for-loop and references to arrays. The mid level example shows the exclusion of array references replaced by pointers. It also shows the declaration of certain arrays into specific memories. As well, the built-in function, `loop` is used for a hardware do-loop, and `MULT` is used for a multiply operation. The low level example shows pointers assigned to specific registers. In addition, a manual loop pipelining operation has been done by specifying new variables allocated to register sets. Finally, the assembly level example shows two in-line assembly instructions specifying specific operations and registers. The function interface shows which registers are used to pass values into and out of the in-line assembly block.

The support of lower abstraction levels of coding is a non-negligible compiler development effort, especially when styles are mixed. Lower level constraints must be propagated to all the mapping phases and algorithms used by the compiler. Algorithms are always easier to implement if they have more degrees of freedom; thus, new constraints can sometimes pose difficulties. For example, to mix the high level

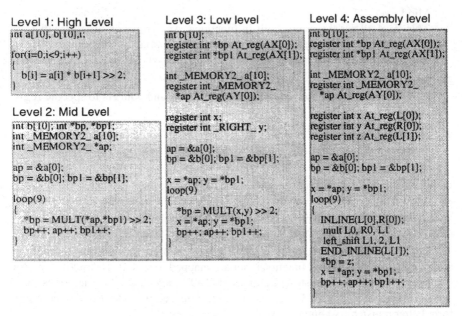

Figure 4.5 Examples of C code at different abstraction levels.

(Level 1.) with the mid level (Level 2.) requires the handling of both pointers and arrays. This means for the inclusion of any array optimization techniques, an *alias-analysis* [1] function is required in the compiler.

4.3 Validation strategies

The very definition of the term *retargetable compiler* suggests a countless number of targets and even targets that have not yet been designed. This places a huge importance on the validation methodology. The confidence that a compiler produces correct code is a significant factor that the embedded system designer cannot neglect.

Compiler validation is done today predominantly based upon simulation. While formal approaches are making in-grounds in the RTL (Register Transfer Level) and logic synthesis areas of hardware synthesis, they lag far behind for any behavioral level specification including C.

For simulation-based validation schemes, selection of a suitable test suite which covers possible faults is an issue which arises. Commercially available C test suites are available, such as Plum-Hall [119], Perenial, and MetaWare [118]. These suites are made up of examples which test all the facilities of C in a thorough manner. Unfortunately the test suites are not directly applicable to embedded processors because an embedded processor compiler typically uses a subset of C. For example, only some data-types (see Section 4.1.1) and some operations may be supported. Moreover, any extensions to C are not tested (see Section 4.2).

The key to a good test suite is organization. A test suite should be organized so that the language support can be parameterized. Therefore, the suite may be *personalized* quickly and efficiently for any new target of the compiler. The available coding levels of the compiler (Section 4.2.2) should be thoroughly tested; however, target specific C extensions such as built-in functions (Section 4.2.1) have to be treated on a case-by-case method.

4.3.1 Instruction-set simulation

While the subject of instruction-set simulation [99][114] is much wider than the mention given here, this section simply provides a definition as it relates to the compiler validation issue.

An instruction-set simulator is an execution model which runs the behavior of the embedded processor, on either the level of operations (instruction-accurate), machine cycles (cycle-accurate), or the netlist (ns-accurate). Any of these levels is sufficient for compiler validation.

Typically, the simulator takes microcode as entry and has interpretive functions which allow the user to run, step, and break the program and to look at register contents and memory. Typical methods used to build a simulator model are: hand-writing in C; or generating from an instruction-set description. From the perspective of validation, both methods are equivalent. However, the latter is naturally favored to reduce the effort in engineering development. The latter may also offer more confidence if the tool has been tested for a large number of processor models.

Having an execution model of the processor is essential to the methodology of validation by simulation. It serves as one side of the balancing scale for the comparison of functionality, as will be described in the next section.

4.3.2 Workstation compilation and bit-true libraries

There are numerous advantages to including a parallel compilation path on the host platform in addition to the target compilation path. The development and debugging environment of the host is generally available prior to the availability of the target development tools. This means that application development can begin before anything else is in place. Even after the target compilation environment is in place, a host equivalent execution will run much faster than any instruction-set simulation of the target processor.

With this methodology in mind, it is important to provide host bit-true functions for:

- any built-in functions that are provided by the target compiler (Section 4.2.1),

- any operators with data-types differing from ANSI C (Section 4.1.1), and

- any other C operations that are implemented differently on the target hardware than on the host processor.

The construction of the bit-true library typically involves careful handling of bit-widths with shifts and bit-masking.

Once a bit-true library is in place, a validation methodology as shown in Figure 4.6 is possible. The function library contains functions which allow writing values to a pre-defined test buffer. After compilation on both paths, this buffer is compared for any differences, which indicate a discrepancy in the retargetable compiler, instruction-set simulator, or the bit-true library. In most cases, the host compiler is assumed to produce correct code, as it usually must pass it's own validation phase.

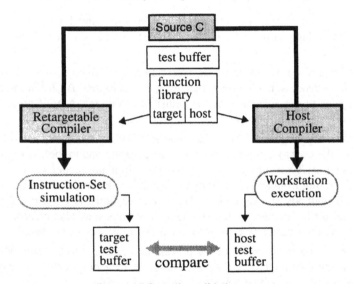

Figure 4.6 Compiler validation strategy

4.3.3 Compiled instruction-set simulation

An alternative approach to compiler validation is *compiled instructions-set simulation*, which, loosely defined, is the interpretation and reconstruction of microcode to be run on another target. One way to achieve this is to *de-compile* the target microcode to a form of C code so that it can be compiled onto the workstation. The result can then be compared to the original C execution on the workstation in the same manner as shown in Figure 4.6.

This methodology has also been used starting from the assembly level, which can provide a faster instruction-set simulation compared to interpretive simulation [114]. However in this case, the source of the compilation is assembly, whereas in validation, it is also important to validate the final assembly step to microcode.

The program which performs the cross-compilation (de-compilation and compilation) can be either handwritten or generated from an instruction-set description. As far as confidence in the validation, using a compiled instruction-set simulation has the same number of steps outlined in the methodology shown in Figure 4.6, as the interpreting instruction-set simulator is either hand-written or generated from an instruction-set description.

4.4 Debugging: how much is really needed?

A compilation path to the host workstation or PC allows standard source-level debugging tools to be used. An example public domain debugger is *gdb*, the GNU source-level debugger distributed by the Free Software Foundation [96], which has many user interfaces (e.g. xxgdb, Emacs, ddd). As the host compilation path is naturally the faster path and the tools are immediately available, functional validation and debugging of the source algorithms should ideally be done at this level.

After functional validation has been done on the host, debugging of the embedded processor may also be needed to debug the validation methodology as shown in Figure 4.6, as problems can arise in the target compiler, instruction-set simulator, or bit-true library. Furthermore, debugging may be necessary for the final system after the chip has been fabricated. Figure 4.7 illustrates a methodology whereby the host source-level debugging interface is reused in different modes. The interaction labelled *mode 1* is the familiar host debugging mode; the interaction labelled *mode 2* is the mode using the instruction-set simulator; and the interaction labelled *mode 3* is the mode which interfaces with a cycle-true model of the processor or the chip itself through an in-circuit emulator (ICE) interface.

Mode 2 of debugging is principally used for verifying the retargetable compiler. It is basically a debugging means for the validation strategy shown in Figure 4.6. Of

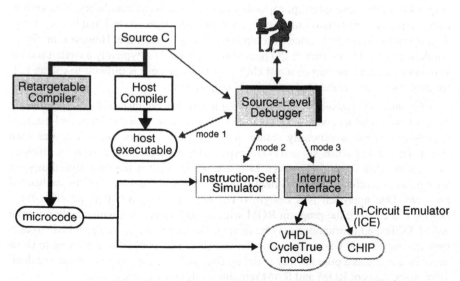

Figure 4.7 Embedded processor source-level debugging.

course, bugs can occur in the instruction-set simulator or the bit-true library as well. In mode 2 of the debugging scenario, depending on the type of instruction-set simulator, the boundary of whether the debugging functions belong to the instruction-set simulator or the source-level debugger can become blurred. For example, general purpose DSPs and MCUs are usually distributed with an instruction-set simulator rich in debugging functions. In this case, the source-level debugger needs simply to provide a user interface to those functions. One method is to detach a user interface (e.g. xxgdb, ddd, Emacs GUD) from a debugger (e.g. gdb, dbx) and attach the user interface directly to the instruction-set simulator. This can often provide a simple route to basic debugging capabilities.

Where the instruction-set simulator does not provide a rich set of functions, it is advantageous to interface the debugger to the instruction-set simulator. Some debugging functions can be quite complex; therefore, depending on the required functionality, this work should not be underestimated. For example, in our experience with the gdb debugger, the memory model has been conceived for VonNeumann type architectures (single memory). The extension to Harvard and multiple memory architectures requires significant re-engineering of the debugger. More efforts on retargetable debuggers are needed [88].

Mode 3 of Figure 4.7 is principally used to debug the hardware. The source-level debugger in mode 3 allows verification of the functionality of the VHDL cycle-true model of the processor. Unless this hardware model can be stopped artificially while retaining its context, interrupt functionality is necessary on the processor. Furthermore, this simulation can be extremely slow since there is an interaction with a hardware simulator (e.g. VHDL) on a detailed level (either RTL or netlist).

The second possibility of interaction in mode 3 is a real-time interface with the chip itself. In this case interrupt capability of the processor is mandatory. This emulation is typically one to two orders of magnitude faster than mode 2 and is even faster than operation in mode 1, since the chip is operating in real-time. However, in-circuit emulation is costly in terms of I/O pins to the exterior. It is typically a route used for verifying a standalone part or a test chip, especially since it is difficult to use an ICE for processor cores embedded on a single-chip system.

An interesting issue related to debugging is the organization of the program memory on a product containing an embedded processor. Since on-chip real-estate is expensive, programs generally reside in ROM since it takes much less area than RAM. This makes sense, since it is not expected that the program need to be changed in an embedded system. However, in test chips it is often the case that things go wrong; and therefore, designers may want to change the contents of the embedded program. One approach is a design to balance this trade-off. It is possible for a designer to enhance the program ROM with a small space of downloadable program RAM. Dedicated instructions may be included in the processor as a provision to execute this *patch* code when debugging the test chip. This method is beginning to show more frequently as a popular way to debug final systems. On the other hand, the decision on the sizes of ROM and RAM remains a difficult guessing game.

Although the mechanics of source-level debugging are well understood, practical implementation can still be a difficulty. The techniques of compilation for embedded processors produce optimized code which is specialized for the architecture. The symbolic debugging of optimized code is an enormous and complex problem. Even the simplest of compilation tasks can produce a tangling mess for the debugging symbols. For example, a register assignment algorithm which assigns the same variable to ten different registers at different points of its lifetime means that the object code must carry ten times the symbolic information than previously. Furthermore, if an aggressive scheduling algorithm is used, operations can move to different points in the code, including into and out of loops. As well, unreachable code optimizations can make code disappear completely.

In addition, older debug formats (e.g. COFF [102][96]) have no way of dealing with optimized code, which has given rise to company-specific variants (e.g. XCOFF, ECOFF, EXCOFF). Some efforts are underway for debugging standards which address some of these concerns (e.g. ELF, DWARF [22][96]). However, for today's embedded processors, the debugging problem for optimized code remains largely unsolved, and users *take what they can get*. That is to say, source-level debugging can work well for some types of optimized code, and not so well for others.

4.5 Chapter summary

This chapter has presented a set of practical design considerations in the establishment of a firmware development environment centered around a compiler for an embedded processor target. First, a number of language issues were considered using C as the source language. Specifically, the areas of data-type support, memory support and procedure calls were discussed. For an application specific architecture, a subset of C data-types are typically chosen and can be interpreted with specialized bit-widths for the target. The support of multiple memories and sections is generally necessary for an embedded processor since the data interface to the peripherals is quite important. Finally, the support of procedure calls requires the consideration of issues such as stack implementation and allocation, as well as interprocedural optimizations.

Next, coding abstraction levels were discussed. It was stressed that performance for real-time systems is critical; and therefore, the support of various levels of coding abstraction is essential for the embedded processor user to reach all the functionality of the chip. Extensions to standard C include built-in functions, storage class allocation, user-register set allocation, and user-register assignment to specific registers.

Following, validation strategies based on simulation were discussed. The essential components of this form of validation include a thorough set of benchmarks which exercise all the parts of the architecture, models for instruction-set simulation, a host compilation path, and a bit-true library. This led to the succeeding topic of source-level debugging. A methodology was outlined which uses an interface connected in three debugging modes. The first is the standard debugging route with the host compiler, the second allows debugging of the target microcode with an instruction-set simulator, and the third allows debugging on a cycle-true model or the chip itself using an ICE. The first mode is typically used to functionally debug the source algorithm, the second mode is used primarily to debug the retargeted compiler, bit-true library, and instruction-set simulator, and the third mode is typically used to debug the design of the processor hardware.

Chapter 5: Compiler Transformations for DSP Address Calculation

This chapter presents a retargetable approach and prototype tool for the analysis of array references and traversals for efficient address calculation for DSPs. Based on a retargetable architecture model, the approach serves as an enhancement to existing compiler systems or as an aid to architecture exploration. This model is a specification of the addressing resources and operations available on the processor which is used to drive the compiler transformations. In addition to providing the transformation for existing architectures, the model allows the designer to tune the operation of the Address Calculation Unit (ACU) toward the application constraints. Variations on the address registers, index registers and hardwired increment and decrement values may be explored for an algorithm by making simple changes to the specification.

5.1 Address calculation units for DSP

The key aspect of a Digital Signal Processor (DSP) is the ability for number crunching. As data intensive algorithms push for higher speeds and throughput, access to data memories becomes the limiting factor. In response to this, designers have conceived the Address Calculation Unit (ACU) (sometimes termed Address Generation Unit (AGU), Address Arithmetic Unit (AAU), Data Address Generator (DAG) or Memory Management Unit (MMU)), an arithmetic unit which works in parallel to the main Data Calculation Unit (DCU). The ACU works solely on address generation to ensure efficient retrieval and storage of data that is calculated on the DCU. In most cases, the ACU works in a post-modify (increment/decrement) fashion to ensure high speed. Pre-modify addressing is rare because this would require at least two operations to occur in the same instruction cycle, namely the address calculation, then the memory access. On a programmer's level, the difference is in the type of supported addressing modes. A post-modify address calculation unit offers the register direct mode with post-operations. Unsupported pre-modify addressing would mean the disappearance of indirect or indexed addressing modes.

Post-increment/decrement address units are present on countless general-purpose DSPs and cores. Some examples include the SGS-Thomson D950 core [91], the Motorola 56xxx series [78], the Texas Instruments TMS320 series [100], the Analog Devices ADSP-21xx series [5] and the Lode DSP Engine [19]. They are also common in Application Specific Instruction-Set Processors (ASIPs), fine-tuned to application areas, such as MPEG audio [12], Dolby decoding [115], and DSP for telecommunications [65].

Let us consider some example post-indexing ACUs for DSPs. Figure 5.1 shows the address calculation unit of the Motorola 56000 series [78]. It contains two identical halves, each with an arithmetic unit which performs post-indexing on separate sets of registers. The two resembling halves of the ACU exist mainly for the two memories X and Y addressed by the address busses XAB and YAB. Therefore, in principle, both halves of the addressing unit may be active in parallel with the central data calculation unit (DCU) and accesses to each of the memories.

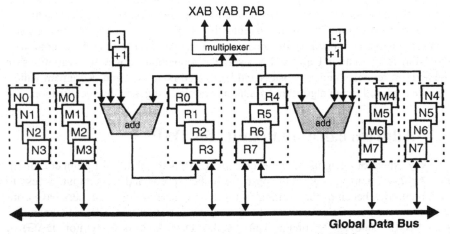

Figure 5.1 Address calculation unit of the Motorola 56K Series

Registers are treated as triplets (i.e. R0 : N0 : M0, R1 : N1 : M1, etc.). An address register, Rn, may be post-incremented only with the index register, Nn, if both registers are within the same triplet. The available operations are summarized in the programming model of Table 5.1. Post-increment and decrement operations are available for the constant 1 or a value within the Nn register. The Mn register determines the type of address arithmetic: linear, modulo, or reverse-carry.

Table 5.1 Programming model for the ACU of the Motorola 56K series

Description of ACU Operation	Uses Mn Modifier	C-like operation	Additional Instruction Cycles
No Update	No	(Rn)	0
Post-increment by 1	Yes	(Rn)++	0
Post-decrement by 1	Yes	(Rn)--	0
Post-increment by Offset Nn	Yes	(Rn)+= Nn	0
Indexed by offset N	Yes	(Rn+N)	1
Pre-decrement by 1	Yes	--(Rn)	1

A second example is the SGS-Thomson D950 Core [91] shown in Figure 5.2. It also contains two halves with separate arithmetic units. In this case, the registers AX0, AX1, and SP address the X data-memory; registers AY0 and AY1 address the Y memory. Post-increment operations may be executed on AX0 and AX1 with any of the index registers IX0, IX1, IX2, and IX3. Similarly, post-increment operations may be executed on AY0 and AY1 with any of IY0, IY1, IY2, and IY3. The SP register can be used for a stack and has special operations such as push (pre-decrement) and pop (post-increment).

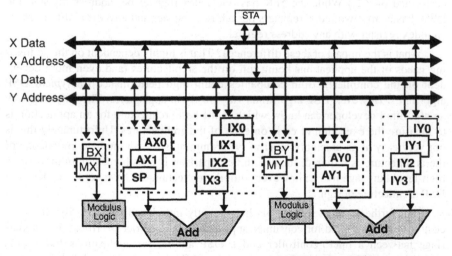

Figure 5.2 Address calculation unit of the SGS-Thomson D950 core

The available operations on the ACU of the D950 are summarized in Table 5.2. Post-increment operations are available for the address registers with index registers within the same half of the ACU. Post-increment and pre-decrement by the constant

1 are available for the SP register. Modulo and bit-reverse addressing is determined by the STA register and the bounds are set in the B and M registers.

Table 5.2 Programming model for the ACU of the SGS-Thomson D950

Description of ACU Operation	modulo addressing	bit-reverse addressing	C-like operation
Post-increment AXn by IXn	Yes	Yes	(AXn)+= IXn
Post-increment AYm by IYm	Yes	No	(AYm)+= IYm
Pre-decrement SP by 1	No	No	--(SP)
Post-increment SP by 1	No	No	(SP)++

Although at first glance, the ACUs of the Motorola 56000 and the SGS-Thomson D950 look very similar they have some very different characteristics. While both are designed for architectures with two data-memories, the 56K ACU allows either of its halves to point to either memory X or Y. The D950 ACU allows each half only to point to its respective memory X or Y. The number and available operations of the address registers differ: the 56K has 8 address registers which can perform post-increment with 1, -1, or its respective index register. The D950 has 4 address registers which can perform post-increment with any index register on its respective side, but not with any constants, with the exception of SP. (Post-increment with 1 or -1 can be performed on SP.) While the 56K has one index register per address register, the D950 has nearly two index registers per address register and also the ability to share an index register with any address register.

What is the impact of these differences? That is highly dependent on the addressing needs of the applications being run on the architectures (and even more dependent on the compiler!). Both companies claim high performance for *typical* DSP algorithms. However, DSP algorithms vary vastly in appearance. The only true way an algorithm developer can know which is the best architecture for an application is to measure the execution of code on each of the architectures. Unfortunately this is not very easy at an assembly-level of programming since an intimate knowledge of the architecture is required. Ideally, the comparison is possible if compiling from a high-level language like C; however compilation techniques have not kept pace with DSP architecture design.

In a further example, Motorola has recently introduced the DSP 56800, a low cost 16-bit DSP geared for consumer applications where price is critical. It is a marriage between a micro-controller and a DSP aiming for applications like digital answering machines, feature phones, modems, AC motor control, and disk drives.

Isolating the ACU of the 56800, one can see that it is based on the functionality of the 56000 ACU, but with variations. Shown in Figure 5.3, the ACU has five address registers, one designed to be used as a stack pointer (SP).

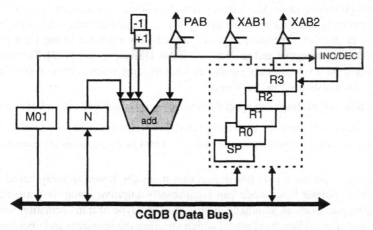

Figure 5.3 Address calculation unit of the Motorola 56800

The available operations of the unit are shown in Table 5.1. Since there is one memory, only one half of the ACU of the 56000 is available. Although the unit has more addressing modes available, there is an instruction-cycle penalty for the indexed modes. The post-modify modes remain the more efficient. Note the restrictions in register uses: only R0 and R1 can perform modulo addressing (bit-reverse is not available); only R2 and SP can perform the short indexed mode.

In addition, register R3 has a special property as shown in Figure 5.3. A post-increment/decrement may be performed in parallel to post-increment/decrements on another address register in addition to up to 2 reads from data-memory.

Table 5.3 Programming model for the ACU of the Motorola 56800 series

Description of ACU Operation	Uses M01 Modifier	C-like operation	Additional Instruction Cycles
No Update	No	(Rn)	0
Post-increment by 1	R0, R1 optionally	(Rn)++	0
Post-decrement by 1	R0, R1 optionally	(Rn)--	0
Post-increment by Offset N	R0, R1 optionally	(Rn)+= N	0
Indexed by offset N	R0, R1 optionally	(Rn+N)	1
Indexed by short (6-bit)	No	(R2+xx) (SP+xx)	1
Indexed by long (16-bit)	R0, R1 optionally	(Rn+xxxx)	2 + extra word

Although these type of ACUs have existed for some time and continue to evolve, the compiler techniques for mapping high-level language constructs onto these register structures are very immature. This is immediately reflected in the poor performance of both commercial and publicly-available DSP compilers [112]. The problem of mapping high-level language structures such as array references onto post-indexing ACUs manifests itself in two ways:

1. difficulties of dealing with special-purpose register connections and operations.

2. difficulties in treating the disjunction in dependency between the use of addresses and the calculation of new addresses, inherent in the post-modify nature of the unit.

Previous experience [64] has shown that manually lowering array-based high-level code to pointer-based code can significantly improve compiler performance. This chapter addresses an automatic approach to this type of transformation with the introduction of an architectural model which specifies the resources and operations of an address calculation unit.

5.2 Traditional address generation techniques

5.2.1 Related work

Approaches to improving the generation of addresses for array references include the work of Joshi and Dhamdhere [52], a strength-reducing technique for induction variables and other loop variables which builds upon the code hoisting techniques pioneered by Morel and Renvoise [77]. These techniques aim at replacing expensive operations such as multiplications with less expensive operations such as additions and subtractions (strength reduction) and moving as much as possible outside of loops (code hoisting). Since it is typical that array references are calculated upon induction variables, transformations on these variables represent the greatest gain. Although these techniques are important techniques for general-purpose processors, the gain for post-modify calculation units is marginal. The techniques are aimed toward pre-calculation of addresses.

Recently, offset assignment techniques have appeared to address post-modify address calculation units [60][63]. These approaches propose the placement of variables in memory based on the use of each variable in the data-flow calculations of the program. This placement is accomplished in a manner which minimizes the number of post-increments on each address pointer. This ensures that the generation of addresses for static variables is optimized. This work does not address the following issues:

1. Optimization of address generation for higher level constructs such as arrays and structures. The majority of DSP programs manipulate arrays and structures for sampling streams, etc. Address generation for these constructs is of fundamental importance in DSP.

2. Optimization of address generation for a design flow which includes separable compilation and linking. As described in Section 4.1.3, memory assignment techniques based on data-flow usage preclude the use of a linking phase common for C compilers. A linking phase is fundamental for libraries of high-performance DSP functions.

Complementary to the work on address generation is the possibility of reordering array indices to improve the use of temporary storage. This is particularly important in video signal processing. The work using the polyhedral dependency graph model introduced by IMEC [29] and the PHIDEO compiler introduced by Philips [72] address these type of transformations. In addition to the reordering of array indices, this work also addresses memory placement which for the same practical reasons as mentioned above can be difficult to use with a linking phase (see Section 4.1.3).

5.2.2 Address calculation for arrays

Address Pre-calculation. A straight-forward method of calculating addresses for arrays is *on-the-fly* generation. For a simple array reference, this involves the addition of a base address with an induction variable (assuming the data size is 1; otherwise a multiplication by a constant is needed) as depicted in the example of Figure 5.4. The shortcoming of this approach is that the value of the address must be calculated before the reference because the operation is data dependent. Within the context of loop bodies, pre-calculations of this sort can have a significant performance penalty.

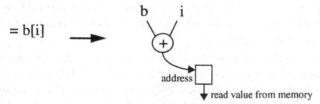

Figure 5.4 Pre-calculation of array addresses

An improvement to this straight-forward approach is loop pipelining [37][55], where addresses for iteration *i* are calculated at iteration *i-1* (see a loop pipelining example in Section 2.6.2). This can be extended depth-wise for the number of operations of the induction variables. For this type of address calculation, the strength reduction and code hoisting work mentioned in Section 5.2.1 is also of use. However, it is important to note that even with these improvements the incremented *values* of the induction variables must still be calculated for each iteration of the loop (usually at the end of the loop).

While being an improvement over the straight-forward pre-calculation of addresses, this type of loop pipeline does not offer a natural mapping to a post-modify address calculation unit, especially within the context of the address register connections. For example, calculations for induction variables of integer type would normally be done on the DCU and would need to be transferred to the ACU for address calculations.

Different techniques are needed for post-modify address generation.

Address Post-calculation. A second approach to address generation involves reducing an array reference to an address (or pointer) reference and incrementing/decrementing the resulting address for the next reference of the array. This optimization has a two-fold advantage:

1. The address calculation can be done in parallel to any principal operations.

2. There is no more need for an induction variable or induction variable calculation (assuming it is not needed for other purposes).

On many DSPs, an available zero-overhead hardware do-loop can perform the loop function. An example is shown in Figure 5.5, where for instance, the next use of the pointer bp is one position higher than the current position of bp.

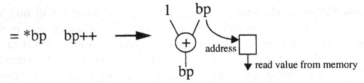

Figure 5.5 Post-calculation of array addresses

This approach maps most naturally to the ACU post-incrementing structure described in Section 5.1. The approach requires, as is also true for loop pipelining, a careful semantic analysis of the subscript dependencies through the various control constructs of the source program.

Figure 5.6 shows an example of the compilation of the code for a simple loop with one array reference before and after a pointer transformation to a post-calculation style of array addresses. The assembly code corresponds to a fictional load/store instruction-set architecture with a hardware do-loop and a simple post-increment ACU which runs in parallel. Note the difference in code size of each example. In addition, the assembly code after the pointer transformation will clearly run faster, as there are only 2 instructions in the loop body as opposed to 6.

For this example in Figure 5.6, it would also be possible to further improve the transformed microcode by performing software pipelining on the loop (see Section 2.6.2), if, for example, loading a register from memory in parallel to other operations were possible on the architecture (R3 <- ld AR). The load could occur once before the loop body (the prologue) and in parallel inside the loop. The loop body could be reduced to one highly compacted line of microcode.

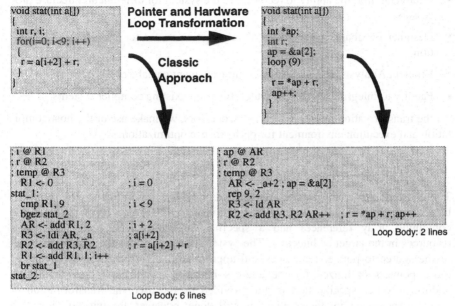

Figure 5.6 Compilation before and after pointer and hardware loop transformation.

5.3 Address transformations for post-modify address calculation

The embedded DSP system is by definition a closed system, responding only to real-time stimuli. Firmware is compiled separately on a desk-top host before residing on the system. In a typical embedded system development environment, this firmware is well simulated and validated before ever reaching the real system. When in the field, the DSP program responds solely to waveform signals through interfaces to its external world such as memory-mapped I/O (MMIO), shared memory, or interface peripherals. We summarize these two important properties of a embedded DSP system as follows:

- a thorough simulation step on the host before the running of the system in field.

- a well-defined boundary between program variables and data variables.

These are two properties which set an embedded DSP system apart from general computing applications. Consequently, they also provide opportunities for new approaches to compilation.

For an embedded DSP system, we propose a transformation of a program's address calculation model which is based on array to pointer optimizations, with the following goals:

- Retargetability: the ability to reconfigure the system for different architecture processors.

- Designer Feedback: a maximum of information useful for architecture exploration.

- Efficient Analysis: a minimum of complex semantic analysis.

- Facility to Integrate: an ease of integrating into existing compiler systems.

As the transformation targets embedded processors, we make use of the host compilation and execution environment for performance optimization.

5.3.1 Overall flow

The complete array analysis and transformation is depicted in Figure 5.7. From the user's viewpoint, only the shaded boxes are visible. He/she provides a C source file containing array references and a specification file indicating the addressing resources in the target architecture. The system then transforms the array references of the source to pointer references and appropriate increments and decrements of those pointers optimized for the address resource specification provided. If an address resource specification is not provided by the user, the transformation of arrays to pointers (pointer creation) is still done; however, the register allocation (pointer combination) and register assignment phases are skipped. These phases could be done by the target compiler.

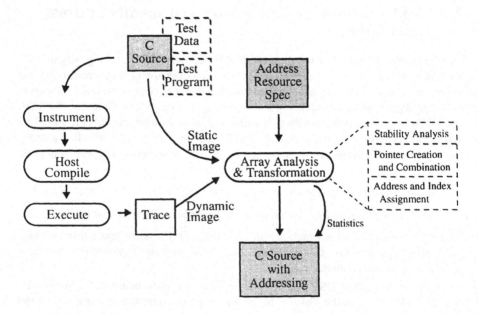

Figure 5.7 ArrSyn array analysis and transformation flow.

In addition, statistics are generated for the user during compilation and as comments embedded in the target (C source with addressing). These statistics include basic block frequencies, array reference frequencies, and the number of pointers created. For the created pointers (of which the number may not correspond directly to the number of arrays), the system also provides the reference frequencies and the frequencies of increment/decrement operations.

The choice of the C language as the target provides the following benefits:

- the target can be compiled and verified against the behavior of the source.

- the target can be used as an input to a dedicated architecture compiler.

- the semantics are easily understood by a human reader.

However, a drawback of using C is that fine-tuning for parallelism is not possible. Parallelization (compaction) is left for the target architecture compiler.

The central analysis block uses both a static and dynamic (run-time) image of the source algorithm. The advantages of using static and dynamic information versus only static information are:

- the ability to determine non-obvious linear relationships.

- the availability of relative frequencies of basic blocks, which indicate a realistic performance cost of the insertion of instructions.

The dynamic image of the source is created through instrumentation of the original source, compilation, and execution on the host. Details are provided in Section 5.3.3. As a result of this methodology, the following items are required:

- the source must be compilable and executable on the workstation.

- a test program must be provided that exercises all the basic blocks to be transformed.

- execution on the workstation should have reasonable run-time.

- induction variables may not be dependent on test data (i.e. data variables which may change from execution on the host to execution in the field).

In our experience, these items are common in an embedded systems development methodology, where firmware is simulated on a desk-top platform before being used in the field. This differs in nature from a general computing environment. For this last item, since the I/O of an embedded program is well defined and contained, induction variable dependency can be easily identified.

5.3.2 *Address resource specification*

An example resource specification is shown in the left side of Figure 5.8. The specification includes two main parts: a declaration of resources (address and index registers) and the operations that can be performed on these resources. This specification is represented internally as a structural connection of registers, adders, and constants.

ACU Specification   ACU Internal Structure

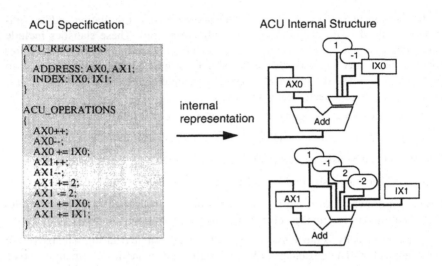

Figure 5.8 Address calculation unit specification and representation

ACU Internal Structure

ACU Specification

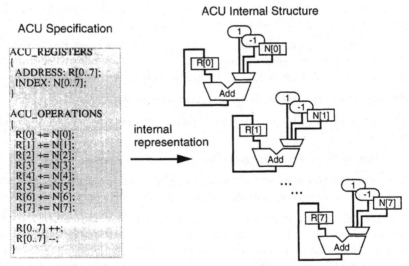

Figure 5.9 Address resource specification for the Motorola 56000

This behavioral representation describes naturally the full operation of the unit and it is useful for allocation and assignment, where pointers and increments can be bound to registers and constants.

A second example is shown in Figure 5.9 for the linear post-addressing portion of the Motorola 56000 DSP (see ACU diagram in Figure 5.1). The specification shows in a very simple manner the operations and resources of the functionality of the address calculation unit. This specification can be easily manipulated by a designer of the unit.

Further extensions to the specification model would include an identification of the functional units on which each operation is performed, as well as any encoding restrictions which are imposed by the instruction-set word. These considerations would naturally overlap with information needed by the target compiler for other phases of compilation including assembly, and would therefore fit it in best with a full instruction-set architecture model.

5.3.3 Instrumentation and tracing

In this context, instrumentation is defined as the transformation of the original source code to a duplicate plus the addition of tags. Tagging is formulated as a lexical and semantic analysis of the source program for the annotation of output statements that indicate run-time information. The tags include: function entries, function exits, loop entries, loop begins, loop-exits, array references including induction variables and run-time *values* of induction variables. All other code is ignored. The run-time values of the array induction variables is the key component which allows the analysis of the array traversals.

Execution of the instrumented code produces a trace that is consumed by the main analysis block of the system. In this manner, the array access patterns can be determined quickly and with a minimum of semantic analysis. Note that tracing has also been used in other contexts to improve run-time performance [58]. The most important point about using tracing is that improvements to the code can be done *exactly* in the places where it is needed: the blocks of code which execute the most frequently.

5.3.4 Stability analysis

Given a reference to an array at a static position in the source program, stability analysis determines if this reference is visited in a linear fashion within a set of loops throughout the execution of a program. If so, it may be replaced by a pointer and an increment/decrement set. One could imagine two forms of stability analysis: static and dynamic. Each has its set of advantages and disadvantages. A stability analysis based on the static image of the program guarantees that the transformation is valid independent of the test data; however, the current state-of-the-art techniques are restricted to loops of very well-defined behaviors, for example loops whose boundaries are calculated by linear affine functions [29]. A stability analysis based on the dynamic image of a program has the advantage of the capability to analyze much more general loop structures, which contain non-obvious linear traversals of array references. Complicated array reference calculations can be determined to be stable, simply by examining the reference progression. However, the stability will only be true for a certain test program. If the test program does not exercise the entire program as it will run in the field, this stability analysis may not hold. Since embedded systems are usually simulated thoroughly on a desk-top before being downloaded onto the chip, a program not exercising all the loops would be a rare case.

The dynamic stability analysis has been implemented for the ArrSyn transformation. The analysis makes use of the dynamic trace of the program which quickly evaluates the characteristics of the array reference run-time progression. Induction variables may not be dependent on test data (values that are changed from execution on the desktop to execution in the field). Thus, a safe methodology is that the user ensures that no induction variables are dependent on data that resides on data memories, memory-mapped I/O locations, or interface registers.

For the dynamic stability analysis, an array reference A can be determined to be stable using the dynamic image produced by tracing. An array reference A is said to be stable within an inner-most loop IL, if the *stride* of A remains constant between every loop begin of IL without crossing a loop entry or exit of IL. The stride of an array reference A, is the difference in the value of the induction variable from one reference to the next. An array reference A is said to be stable within a loop L if it is stable for all its encompassing loops including the innermost loop. Array references which are stable within a loop may be eventually replaced by a pointer reference and a set of increment/decrement operations or combined with another pointer reference as described next.

5.3.5 Pointer creation and combination

Pointer creation and combination is the allocation phase of the analysis. The goal is to produce an appropriate number of pointers which match the capabilities of the address calculation unit. The approach begins by creating a pointer for each static array reference of the source program, given that the array reference is stable within a set of encompassing loops. This starting point uses the maximum number of addressing resources, one pointer for each array reference. From this point, static pointers are combined. Two references are made to use the same pointer through correctness preserving combinations. This is done until a reasonable number of pointers exist for the architecture at hand.

The combination strategy uses the following combining rules:

- pointers created for array references with exactly the same *signature* within the same nest of loops may be combined. The signature of an array reference corresponds to the programmer's view of the elements in the array. (e.g. b[i+2] and b[i+2] have the same signature, whereas b[i] does not).

- pointers with non-overlapping lifetimes may be combined.

- pointers referencing the same array at different relative positions and progressing in the same fashion within the same nest of loops may be combined.

As these transformations have various effects on the resulting code, rules are executed with the following objective functions:

- reduce the number of pointers to an amount equal or below the number of available address registers.

```
for (i...              for (i...              for (i...              for (i...
{                      {                      {                      {
    while (j...            while (j...            while (j...            while (j...
    {                     {                     {                     {
        b[i] = b[i] + ...     *b_1 = *b_2 + ...     *b_1 = *b_1 + ...       *b_1 =* b_1 + ...
        b[i+1] = ...         *b_3 = ...            *b_3 = ...              b_1++ ;
        a[j] =...            *a_1 =...             *a_1 =...               *b_1 = ...
    }                        a_1++;               a_1++;                  b_1--;
    b[i] = ...            }                     }                        *a_1 =...
}                        *b_4 = ...            *b_1 = ...               a_1++;
do                       b_1++;                b_1++;                }
{                        b_2++;                b_3++;                *b_1 = ...;
    x[j] =               b_3++;             }                        b_1++;
} while (j...            b_4++;             do                       }
                      }                     {                        do
                      do                        *b_1 =                {
                      {                         b_1++;                    *b_1 =
                          *x_1 =            } while (j...                 b_1++;
                          x_1++;                                     } while (j...
                      } while (j...
```

a) Original Code	b) Pointer Creation	c) Pointer Combination	d) Pointer Combination

Figure 5.10 Pointer creation and combination example

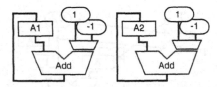

Figure 5.11 Example target ACU internal structure

- minimize the frequency of inserted increments/decrements of pointers.

- minimize the number of different valued increment/decrements for each pointer.

Figure 5.10 shows an example of the pointer creation and combining process. Consider a target ACU with 2 address registers, such as the one with the internal representation shown in Figure 5.11. Figure 5.10 a) displays the original C code; Figure 5.10 b) displays the creation of pointers, one for each array reference, a total of 6 pointers (b_1, b_2, b_3, b_4, a_1 and x_1) (Initializations of pointers are not shown so as to simplify the figure); Figure 5.10 c) shows a first application of pointer combinations, the array references with exactly the same signatures and those with non-overlapping lifetimes. This results in a reduction to 3 pointers (b_1, b_3 and a_1). Figure 5.10 d) shows a possible second application of pointer combinations (b[] at different relative positions), reducing the number to 2 pointers (b_1 and a_1). This number of applications of pointer combinations depends on the availability of address registers in the ACU specification. For this example, the goal was to reduce the number of pointers to at most 2, to correspond to the number of address registers in the specification.

Notice that after pointer creation, the loop induction variables (i and j) are no longer needed for referencing. Thus, the loops which use these induction variables may subsequently be mapped to any available hardware do-loops.

5.3.6 Address and index register assignment

Following their creation is the assignment of pointers and increments to address registers and index registers or constants (hardwired constant values). Lifetime analysis has already been done in the combination stage; therefore, the formulation of the problem is to find the best one-to-one matching of pointers to address registers and their respective increments to constants or index registers. If the number of pointers that exist after combining is less than the number of address registers, a direct mapping is usually possible. If the number of pointers happens to exceed the number of address registers, then some pointers must be assigned to memory and will be stored and loaded into a free address register.

Before we explain the assignment strategy, we shall define some terms. A pointer, p, is said to be *fully assigned* to an address register, A, when p is assigned to A and the increment/decrement values associated with p are assigned to index registers of A, (I_A) and/or constants of A, (C_A). This is depicted graphically in Figure 5.12.

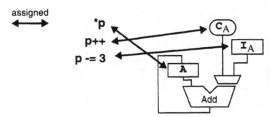

Figure 5.12 Pointer p fully assigned to address register A

We define the *reference occurrence* (RO) of a pointer p to be the number of times a value in memory is referenced by p in the execution of the program.

We define two objective cost functions for a pointer which is fully assigned to an address register. For a fully assigned pointer:

1.   the *best assignment cost* (BC) is defined as the number of address calculation instruction executions in the final code.

2.   the *estimated assignment cost* (EC) is defined as the best cost weighted by the probability that the indexing resources (i.e. index registers) be free for use.

The cost for an assignment is based on the frequency of increment and decrement instructions in the final code. This best assignment cost function reflects the number of times an ACU operation will eventually be executed in the final code.

These 3 values, RO, BC, and EC can be calculated by making use of the dynamic information provided by tracing, namely the frequency execution of basic blocks. The best assignment cost function (BC) does not correspond directly to the number of whole instructions that will be executed in the final code, since on most architectures many of these instructions may be compacted in parallel with other operations. However, it is a good reflection of the trade-off between different assignments. Similarly, the reference occurrence (RO) reflects the number of times a pointer will be used to store or retrieve data in the final code.

Two fully assigned examples are shown in Figure 5.13. For the pointer bp which is fully assigned to AX0. The assignment to the address register AX0 and constant indices of +1 and -1 results in a best assignment cost (BC) of 17, since an initialization AX0 = &b[n] will be executed once, the operation AX0++ will be executed 12 times, and the operation AX0-- will be executed 4 times. For this example we assumed that the initialization occurs once; the actual number of initializations and the value of the constant n are both dependent on the context in the program.

For the pointer xp in Figure 5.13 which is fully assigned to AX1 (address register AX1, the constant +2, and the index register IX0) gives a best assignment cost (BC) of 66, since in the final code the initialization AX1 = &x[n] will be executed

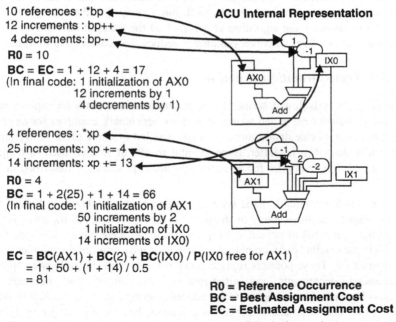

10 references : *bp
12 increments : bp++
4 decrements: bp--

RO = 10

BC = EC = 1 + 12 + 4 = 17
(In final code: 1 initialization of AX0
                12 increments by 1
                4 decrements by 1)

4 references : *xp
25 increments: xp += 4
14 increments: xp += 13

RO = 4
BC = 1 + 2(25) + 1 + 14 = 66
(In final code: 1 initialization of AX1
                50 increments by 2
                1 initialization of IX0
                14 increments of IX0)

ACU Internal Representation

EC = BC(AX1) + BC(2) + BC(IX0) / P(IX0 free for AX1)
   = 1 + 50 + (1 + 14) / 0.5
   = 81

RO = Reference Occurrence
BC = Best Assignment Cost
EC = Estimated Assignment Cost

Figure 5.13 Examples of the reference occurence (RO), the best assignment cost (BC) and the estimated assignment cost (EC).

once, the operation AX1 += 2 will be executed 50 times, the initialization IX0 = 13 will be executed once, and the operation AX1 += IX0 will be executed 14 times. The estimated assignment cost (EC) is calculated as the best cost formula weighted by the probability that the index register IX0 be free for use for AX1. The probability P(IX0 *free for* AX1) is 0.5 because IX0 may be equally free for use with either AX0 or AX1. The estimated assignment cost (EC) is 81, which reflects the fact that it is may be more costly to use IX0 since it may not be free for use by AX1.

The assignment procedure is divided into 2 phases:

1. Determine a direct mapping feasibility: If the number of pointers is less than or equal to the number of available address registers (i.e. the pointer combination phase has succeeded), proceed to step 2. If not, assign the pointer with the smallest RO to memory. Repeat until the number of pointers is equal to the number of address registers.

2. For the direct mapping, fully assign pointers to address registers.

The assignment strategy of step 2 is subdivided into the following steps:

1. Exhaustively determine the lowest EC for each pointer supposing it were fully assigned to each address register.

2. Take the cheapest EC and execute the corresponding full assignment.

3. Update all the estimated costs (ECs) based on this assignment. Repeat step 2.

This estimated cost guides the assignment heuristic since it can determine the best places for potential savings. As well, during step 3 of the strategy, if an index register is assigned, the algorithm keeps track of the value. It later attempts to share common index values wherever possible to reduce the number of initializations.

5.3.7 Transformation example

The array analysis flow described in the previous section has been implemented in a prototype called *ArrSyn* and tested on various benchmark examples for existing and possible address calculation units. A small detailed example of an ArrSyn transformation is described hereafter. Experimental results of ArrSyn used in conjunction with a dedicated compiler for a multimedia audio processor is described in Section 6.4.

A detailed example of the transformation process is shown in Figure 5.14. A close inspection shows many of the features of the system. All the array references have been changed to pointer references. Although there were 8 static array references in the original code, combination strategies have reduced the number of created pointers to 3. These pointers replace array references in the code with appropriate increments, decrements, and initializations. The pointers (and increments/decrements) have also been assigned to the address registers (AX0, AX1, AX2), index register (IX), and constants (+1, -1) in a manner best fitting the given architecture specification. The assignment of pointers to registers may be passed to a dedicated compiler using C extensions as described in Section 4.2.2.

C Source

```
#define N 6

int r;
int a[N] = {2,3,12,6,14,18};
int b[N] = {2,1,9,83,5,-98};

void main()
{
  register int i,j;

  r = 0;
  for(i=0; i<N/2; i++)
    {
      for(j=2; j>=0;j--)
      {
        a[i*2+j] = b[i] + 7;
      }
      b[i] = b[i] + 3;
    }

  j = N-1;
  i = 0;

  while(j--)
    {
      b[j] = a[j+1];
      r = a[i] + b[j] + r;
      if(i == 3) break;
      i++;
    }

  printf("\nr = %d\n",r);
}
```

ArrSyn ➡

ACU Specification

```
ACU_REGISTERS
{
  ADDRESS: AX0, AX1, AX2;
  INDEX: IX;
}

ACU_OPERATIONS
{
  AX0++;
  AX0--;
  AX0 += IX;

  AX1++;
  AX1--;

  AX2++;
}
```

C with Addressing

```
#define N 6

int r;
int a[N] = {2,3,12,6,14,18};
int b[N] = {2,1,9,83,5,-98};

void main()
{
/* ArrSyn : refs: (23x) incs: +1(3x) -1(4x) */ register int *b_AX1;
/* ArrSyn : refs: (13x) incs: +5(3x) -1(13x) */ register int *a_AX0;
/* ArrSyn : refs: (4x) incs: +1(4x) */ register int *a_AX2;
/* ArrSyn */ register int IX;

  register int i,j;

  r = 0;
/* ArrSyn */ b_AX1 = &b[0];
/* ArrSyn */ a_AX0 = &a[2];
/* ArrSyn */ IX = 5;
  for(i=0; i<N/2; i++)
    {
/* ArrSyn : Loop executed 3 times. */
      for(j=2; j>=0;j--)
      {
/* ArrSyn : Loop executed 9 times. */
/* Original :    a[i*2+j] = b[i] + 7; */
/* ArrSyn */ *a_AX0 = *b_AX1 + 7;
/* ArrSyn */ a_AX0--;
      }
/* Original :      b[i] = b[i] + 3; */
/* ArrSyn */     *b_AX1 = *b_AX1 + 3;
/* ArrSyn */ b_AX1++;
/* ArrSyn */ a_AX0 += IX;
    }

  j = N-1;
  i = 0;

/* ArrSyn */ b_AX1 = &b[4];
/* ArrSyn */ a_AX0 = &a[5];
/* ArrSyn */ a_AX2 = &a[0];
  while(j--)
    {
/* ArrSyn : Loop executed 4 times. */
/* Original :      b[j] = a[j+1]; */
/* ArrSyn */     *b_AX1 = *a_AX0;
/* ArrSyn */ a_AX0--;
/* Original :      r = a[i] + b[j] + r; */
/* ArrSyn */     r = *a_AX2 + *b_AX1 + r;
/* ArrSyn */ b_AX1--;
/* ArrSyn */ a_AX2++;
      if(i == 3) break;
      i++;
    }

  printf("\nr = %d\n",r);
}
```

Figure 5.14 Example transformation: C source to C with addressing

In addition to the rewrite of code, the designer is given statistics: profiling frequencies of the execution of blocks, the number of occurrences of references of each created pointer and the number of occurrences of increments of different values (top-right of Figure 5.14). In addition, other statistics are printed at compile time, such as the frequency of reads and writes of the original arrays.

Note that after the replacement of arrays with pointer references, the loop counters (i and j) are no longer used within the two nested *for* loops. These loops are ideally suited to be mapped directly to zero-overhead hardware do-loops (if they exist in the architecture). A back-end process could recognize this in a control/data-flow analysis and produce the correct mapping. The *while* loop can also be replaced by a hardware do-loop, but care must be taken in the control/data-flow analysis because there are two possible exit points.

5.4 Summary and future work

The contribution of this chapter has been to introduce an approach to transforming C code which makes efficient use of address calculation units (ACUs) on embedded DSP architectures. The strength of the approach is a specification model which describes the resources and operations of the architecture. This model allows the designer to evaluate applications on different ACU architectures. A rapid evaluation can be performed by simple changes to the specification model. The output of the transformation is C code with explicit pointer addressing. The advantage of C code as the target is that the output may be fed to any processor-specific C compiler. Furthermore, the semantics of the output are easily understood by the programmer.

The main analysis portion of the transformation makes use of a dynamic trace from a host execution of the program. The ability to do this is a feature which sets the embedded application apart from a general computing application. Moreover, it is this feature which provides the fuel to the transformation algorithms to optimize the most critical portions of the code, which is clearly an advantage over static methods. The approach could be called *profiler-driven*, as the run-time is improved only after a host execution of the code.

The compilation approach has been implemented in a prototype tool called *ArrSyn* and tested on benchmark examples (discussed in Section 6.4).

Future work includes improvements to the combination algorithm to handle architectures with very few resources. Practical extensions to the tool include the provision for different sized data-types and the handling of multi-dimensional arrays and structures, which are important in areas such as video processing.

On the side of the specification model, common DSP Address Calculation Unit features such as modulo and bit-reverse addressing would be a significant improvement to the approach. As well, efforts to merge the specification with a complete model of the architecture sufficient for the entire compilation process would be welcome.

Chapter 6: Pushing the Capabilities of Compiler Methodologies in Industry

For DSPs and ASIPs of many types, assembly level programming is common place; and therefore, experiences with compilation methodologies is an important step for the general acceptance of embedded software tools for specialized architectures. This chapter presents industrial experiences in three different projects using the two compiler systems described in Chapter 3. In each section, the architectures and compiler environments are presented followed by a discussion of the results and lessons learned. These lessons also include many of the practical issues discussed in Chapter 4. Following, experimental results of the DSP address generation approach of Chapter 5 are presented. The chapter concludes with a discussion of the advantages and disadvantages of the main principles of each compiler approach.

6.1 A Nortel ASIP for telecommunications

6.1.1 Architecture description

In Figure 6.1, a diagram is shown for a DSP which was developed in-house at Bell-Northern Research/Nortel. This Application Specific Instruction-Set Processor (ASIP) was developed for a private local telephone switch called a key system unit.

The architecture is inspired by VLIW principles. It is a Harvard, RISC architecture containing an ALU, Multiply-addition unit, ACU, and control unit. The 40 bit instruction word allows a significant amount of parallelism supporting: a control-flow operation, ALU operation, immediate, load-from or store-to memory, and an address calculation operation. The architecture has a number of features which set it apart from other DSPs. Bus connections have been reduced by making specific connections to registers, thereby reducing the instruction decoding but also the homogeneity of the register files. There is one register, R1, which can store data from the ALU to memory; there is one register R6, which may be used to move a calculated address on the ALU to the address register, AR. One register, R7, may be used to hold an immediate value coming from the instruction word. Finally, the register R0 is a constant zero source and bottomless sink.

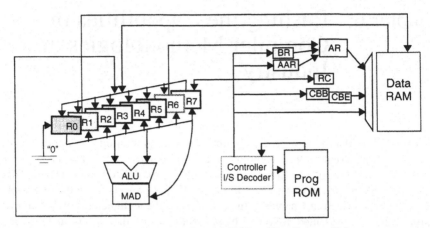

Figure 6.1 Nortel ASIP for a Key System Unit.

These measures place challenges on the development of software tools; however, they allow two significant architectural gains: speed through the reduction of multiplexers and shared busses, and fewer encoding combinations of the instruction word. The latter impacts directly on the needed instruction width; and consequently, the size and expense of the program ROM.

The ALU contains two barrel shifters, one at an input to the ALU and the other at the output of the ALU. This allows the support of various data-types which can be scaled at any moment without instruction delay penalty. Arithmetic instructions can be coupled with input and output shift instructions.

A post-modification address calculation unit is available for parallel execution. Addresses may be computed on the Auxiliary Address Registers (AAR) as well as the standard Address Register (AR). A Base Register (BR) is available for offsets in a custom addressing mode. A circular buffer mode is also available by means of the registers CBB (Circular Buffer Begin) and CBE (Circular Buffer End).

In addition to standard control-flow constructs like conditional/unconditional branches and subroutine calls, one level of hardware do-loop is available.

6.1.2 Compiler environment

As this is a custom architecture, no compiler had originally existed. Attempts to reconfigure a commercially available compiler was successful only with a few case tests. The compiler failed with the large majority of source C tests: some which were written expressly for the architecture, some which were general routines from publicly available DSP sources. The difficulty in reconfiguring the commercial compiler stemmed mainly from the number of special-purpose registers which overlap with the data calculation registers. The solution in this case was to reserve all of the special-purpose registers, which means the compiler quickly runs out of data calculation registers for general arithmetic, hence the inability to compile many of the sources.

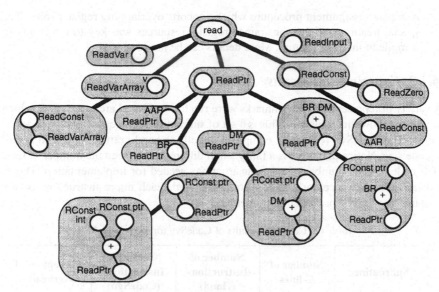

Figure 6.2 CodeSyn data-flow prune tree below the *read* node for the Nortel ASIP

While the model-based CodeSyn compiler described in Chapter 3 was designed to be a generally retargetable compiler for ASIPs, this Nortel ASIP was used to drive the development of the methodology. The pattern base of the compiler was collected directly from the specification in the programmers manual. Figure 6.2 shows an excerpt of the generated pattern tree for the *read* portion of the prune tree. The pattern set contains a certain number of patterns for specializations of the architecture, for example: address generation requires annotations of certain register classes (AAR, BR, DM (DataToMemAddr)) on the input and output terminals of some patterns, and a ReadInput pattern is used for memory-mapped I/O. Although specialized, these branches have the same prune tree organization described in Section 3.2.3 which permits efficient pattern matching.

The full data-flow pattern set for the Nortel ASIP contains the pattern branches of Figure 3.7, the read portion of Figure 6.2, a *write* portion similar to the read portion, and an extra set of address calculation patterns. A control-flow pattern set rounds out the full pattern set needed for the compiler.

The model-based approach brought two distinctive contributions which allowed compilation for the Nortel processor:

1. A pattern matching mechanism which allows the combination of multiply-accumulate and arithmetic/shift operations with no limitations on pattern-size. In addition the patterns support the concept of register classes and are organized in an efficient matching organization. More details can be found in [65].

2. A register assignment procedure which supports overlapping register roles. The special treatment of the constrained register resources was key to the ability to compile to this architecture. More details can be found in [66].

6.1.3 Experimental results

A variety of representative benchmarks were run for the Nortel architecture, of which a subset is shown in Table 6.4. This is a set of subroutines for a conference-call application. The CodeSyn results are compared directly to hand-crafted code with regard to code size. Notice that there is a fairly small difference between the number of lines of C code and the number of micro-instructions needed for implementation. This is indicative of the processor's large instruction word. Each micro-instruction usually contains two to four parallel micro-operations.

Table 6.4 Code size results of CodeSyn for the Nortel ASIP

Subroutine	Number of C-lines	Number of Instructions (Hand)	Number of Instructions (CodeSyn)	Percentage Overhead
Loudest	16	16	21	+ 31%
selectLoudest	39	65	75	+ 15%
selectSpeakers	23	21	36	+ 71%
linearize	11	15	19	+ 26%
sum	14	17	12	- 29%
compander	22	23	25	+ 9%
Overall	125	157	188	+19%

For this very specialized architecture, the benchmarks are an important indication that a compilation methodology can succeed even for processors with highly constrained register files. These initial results do not show superiority over hand coding in assembly; however, as the compiler contains few optimization passes, it does show promise for the approach.

The rest of the examples that were run on this architecture are not reported here mostly because hand code was not written for the equivalent C sources. The main message was that compilation was successful for all of the examples thanks to the treatment of overlapping register classes.

6.2 The SGS-Thomson Microelectronics Integrated Video Telephone

The SGS-Thomson series of videophone systems (e.g. STi1100 [92]) is an example of a system-on-a-chip which contains a set of operators which communicate through a set of busses. The block diagram is shown in Figure 6.3.

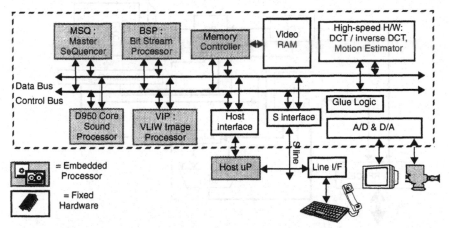

Figure 6.3 The SGS-Thomson single-chip Integrated Video Telephone

Some of these operators are designed as fully hardwired blocks to meet the performance requirements. In this case, behavioral synthesis methodologies are important for hardwired blocks (Architectural synthesis for the Motion Estimator of the IVT is discussed in [14]); however, to keep pace with the evolving standards, many of the IVT operators are designed as ASIPs [39] (Application Specific Instruction-Set Processors). With the constantly changing standards (e.g. H.261, H.263), block functions in software allow for late design changes and modifications, which are important in meeting the current market requirements.

With a wide variety of processor operators, the requirements for the compiler system is that it be easily retargetable to control-flow dominated architectures as well as to data-flow dominated architectures. It must adhere to strict hardware requirements like bus interface protocols as well as handle architecture specialization. Furthermore, since the system is real-time reactive, a performance overhead with respect to hand code cannot be tolerated.

Compilers were developed for three embedded processors of the videophone system shown in Figure 6.3: the MSQ (MicroSeQuencer), the BSP (Bit-Stream Processor), and the VIP (VLIW Image Processor). The D950 core is also sold as a standalone part and already has a C compiler. The memory controller, is a highly specialized block for memory-handling and has its own dedicated compilation utility.

6.2.1 Architecture descriptions

Figure 6.4 shows the architecture of the MSQ (Micro-SeQuencer), which is the top-level control unit of the videophone system. The architecture is a single execution stream controller providing standard ALU operations (ADD, SUB, AND, OR, CMP, SHIFT); as well as standard control operations (BR conditional/unconditional, BR indirect). Reserved instructions perform the function of the bus interface protocol. A

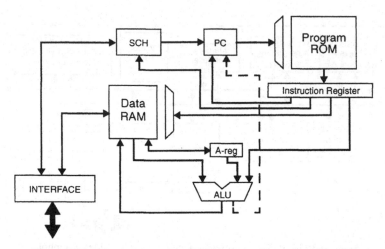

Figure 6.4 MSQ controller of the SGS-Thomson Integrated Video Telephone.

unique property of this block is a unit known as the scheduler (SCH) which can affect the position of the program counter independent of the natural execution order of the program. The scheduler can access the interface directly and make decisions depending on values from the exterior.

The Bit Stream Processor (BSP) is a processor used mainly for the variable length decoding of macroblocks. As one part of the videotelephony decoding process requires many intricate bit-manipulations, the ALU of the BSP can perform a number of bit-level functions. Apart from this feature, the BSP shares a similar architecture template with the MSQ including the bus interface and excluding the scheduler.

The VLIW Image Processor (VIP) is an embedded processor used for prediction routines. The highly-parallel architecture allows the video telephone to do the advanced motion compensation suggested in the H.263 standard. The architecture again shares the same bus interface as the MSQ and BSP; however, in contrast there are several functional units which work in parallel and allow a high throughput of calculations.

6.2.2 Compiler environment

The rule-driven compilation approach described in Section 3.3 was used to develop the compilers for the Integrated Video Telephone. For each of the architectures, a functional rule base was typically developed in two person weeks, or roughly half of the total targeting time. This allows early feedback to the architecture design team before the final refinements are made. Each compiler supports a subset of C; however, support of the entire functionality of the architecture is always available.

For the MSQ, the mapping of the C source to standard arithmetic and control operations was relatively straightforward. The architecture supports only one data bit-width and therefore only one C data-type is supported. Issues that arise are in the routing of data to the special A-register and appropriate scratch locations in the RAM. This is easily handled in virtual code selection. Architecture specific features required special attention, such as the mapping of case statements onto the indirect branching instruction, which requires alignment upon specific bits. This is handled in the mapping to the target machine, where a rule emits an alignment directive along with the assembly code. Other similar features exist and were handled just as easily with the appropriate rules.

For the interface, volatile register variables were defined for use in the hardware operations. In this manner, compiler rules map reads and writes of these variables onto the appropriate special instructions. It is important that the variables are defined as volatile, otherwise compiler optimizations could remove seemingly redundant read and write accesses.

The C compiler for the BSP (Bit Stream Processor) (Figure 6.3) had similar targeting issues to that of the MSQ. The main differing issue was the treatment and optimization of bit manipulation operations. The handling of the register interface was reused from the MSQ.

For the VIP, the declaration of compaction resources was a fundamental development issue, due to the very large instruction word (VLIW) and multiple execution streams. In addition, several built-in functions which correspond directly to hardware functions needed to be designed (see Section 4.2.1). Often, a hardware function of the processor does not correspond directly to C operations. Again, the register interface was reused from the MSQ. In this case, compaction is disallowed with interface functions. This can be guaranteed through a careful definition of the compaction resources (see Section 3.3.4).

6.2.3 Experimental results

Compiler Validation. The validation strategy used for the IVT processor operators is slightly different from the procedure presented in Section 4.3. As the processors of the IVT are an integral part of the entire system, the most important aspect to verify is the function in the system. The validation of the processor functionality was done as shown in Figure 6.5.

The key to the methodology is the presence of a co-simulation approach which allows the simulation of C code on the behavioral level integrated with a VHDL model of the system [67][80][104]. Individual simulations of the system replacing the co-simulation model with a VHDL model of the processor generates two simulation traces which can be compared. The simulation methodology validates both the processor compiler and the VHDL processor model, given a test bench which thoroughly exercises the application C code. Notice that the methodology is similar to the one presented in Section 4.3 since the two principal elements are present: the host compiler and the target compiler.

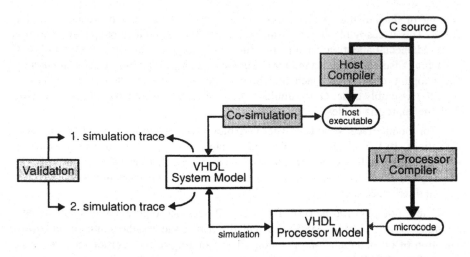

Figure 6.5 Validation of the processors of the SGS-Thomson Integrated Video Telephone

Compiler Results. For the MSQ architecture, a subset of the H.261 code was taken from a prior design of the chip. This code had previously been written in process-level VHDL and was hand-translated to assembly. The examples contain a cross section of the different types of tasks the MSQ performs. This code was rewritten in C nearly line-for-line and compiled using the rule-driven C compiler. We then compared the compiled code with the hand-translated code written in assembler. The results are shown in Table 6.5.

Table 6.5 Code size results of FlexCC for the SGS-Thomson IVT MSQ

H.261 Example	Number of C-lines	Number of Instructions (Hand)	Number of Instructions (FlexCC)	Percentage Overhead
grabber	209	189	203	+ 7%
motion	120	318	311	- 2%
idct_out	293	592	587	- 1%
host_interface	485	710	676	- 5%
Average Overhead				**- 1%**

On average, the compiled code size is roughly equal to the hand written code size. This indicates that for this processor, the compiler performs as well as an assembly-level programmer. This was possible because of the natural mapping from C to the instruction-set of the controller architecture. Only special cases needed to be addressed using the flexibility of rules.

For the BSP and VIP video processors, C compilers were also developed. The compiled code met all code size and performance constraints. Although we have no comparisons with hand code, this is a positive outcome of the previous benchmark which led to a decision by the design team to write all the code in C.

Conclusion. The strength of the rule-driven approach for the operators of the SGS-Thomson Integrated Video Telephone is in its wide flexibility and rapid set-up time. Each of these compilers required roughly one person-month of development. The key lies in the well-bounded functionality of the hardware of each processor. Each programmable operator performs a limited number of tasks. This simplifies the development of the compiler and all the software development tools. At the same time, the performance of the hardware is quite high since it is streamlined for certain operations. However, the performance streamlining is not restrictive since flexibility is still available through the programming of the embedded software.

6.3 The Thomson Consumer Electronic Components Multimedia Audio Processor

The MMDSP multimedia processor was developed at Thomson Consumer Electronic Components (TCEC) for high fidelity audio processing including the decompression and decoding of MPEG2, Dolby AC-3 and Dolby Pro-logic. In addition to being a stand-alone product (the STi4600 [94]), the architecture can also be embedded as a core for integrated products. The processor is used in applications such as Digital Video Disk (DVD), multimedia PC, set top boxes (satellite), High Definition Television (HDTV) and high-end audio equipment.

This project begins with a history of using a well-defined design process for an instruction-set architecture [12]. This methodology includes the use of a special macro-assembler known as *RTL-C*. Source code is written in a form which follows the syntax of C with a number of strict guidelines. Variable names refer to specific registers of the architecture and C operators map directly onto operations in the architecture. For operators that do not exist in C, built-in functions are used (see Section 4.2.1). Each line of C refers to parallel executing operations on the processor.

While this style is restrictive for high-level coding, it has strengths over pure assembly coding, specifically in the architecture refinement phase. The use of the C language syntax allows a second path of compilation on the workstation in order to validate the algorithm behavior. In addition, it allows the use of standard Unix profiling utilities to measure real-time performance before the processor is designed. These utilities include the profiling functions of the cc and gcc compilers and profiling utilities such as tcov, prof, and gprof. More about profiling is discussed in Section 7.5.

While this approach is effective for architecture exploration, when no further changes are made to the hardware design, software development productivity is low since application code must be written on a level comparable to macro-assembly. With the increasing complexity of the MPEG audio standards, the need for an optimizing C compiler had arisen. The requirement for the compiler is that it allow higher productivity by allowing code to be written on a more abstract level and that it not compromise the quality (performance and size) of the code which can be written at the assembly level.

6.3.1 Architecture description

The architecture designed by TCEC is a Harvard, VLIW, load/store instruction-set processor and is shown in Figure 6.6. Communication is centralized through a bus between the major functional units of the ALU (Arithmetic and Logical Unit), ACU (Address Calculation Unit), and memories. The controller is a standard pipelined decoder with the common branching capabilities (jump direct/indirect, call/return), but also including interrupt capability (goto/return-from interrupt) and hardware do-loop capability. Three sets of registers are used to provide three nesting levels of hardware do-loops; however, this can be increased without limit by pushing any of these registers onto the stack.

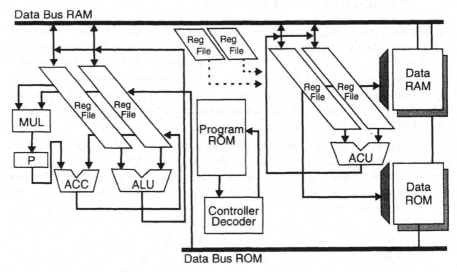

Figure 6.6 Thomson Consumer Electronic Components MMDSP
architecture block diagram

Although the basic design of the unit can be compared to many classic processor architectures, there are certain features which allow it to perform well in this application domain. The post-modify ACU includes custom register connections and increment/decrement capabilities which allow addresses to efficiently traverse the special memory structures. This includes not simply increment by one, but increment and decrement by selected constant values. As well, the increment and decrement values may be held in dedicated registers. One last possibility is the capability to use a constant increment value coming from the instruction register. With this large number of possible operations that can be performed on the ACU, certain combinations are chosen to be encoded in the instruction set such that they execute in parallel to other operations.

The ACU has been designed to work in concert with the memories. The memory structure has been developed around the data-types needed for the application and anticipating future applications to be run on the architecture. A first partition separates memory into ROM mostly for constant filter coefficients, and RAM to hold intermediate data. For each of these memories, several data types are available, some are high precision for DSP routines, others are lower precision mainly for control tasks. In addition to the standard memory locations, there are memory-mapped I/O addresses for communication with the peripherals.

The MAC (multiply-accumulate) unit was designed around the time-critical inner-loop functions of the application. The unit has special register connections which allow it to work efficiently with memory-bus transfers. In addition, certain registers may be coupled to perform double precision arithmetic.

6.3.2 Compiler environment

The rule-driven approach described in Chapter 3 was used to develop the compiler environment. In addition to the retargeting effort of the standard suite of tools, custom optimizations and interfaces were developed to provide a complete, firmware development environment. Some of these are custom optimization modules required for higher performance; many of these are simply tools required to interface into the design environment of the hardware team. The complete environment is shown in Figure 6.7.

Custom Data-type Mapping. For this architecture, the first key item to resolve was the support of the custom memory structure. This memory structure posed a unique challenge which stems from the multiple data-types and memories with varying addressing strides.

Inherently, the retargetable compilation system handles multiple memories and multiple data-types. However, it is required that all memories be of the same bit-width. This implies that data-types of increasing widths take either the same or more memory spaces. For example, if there are three data types $dtype1$, $dtype2$, $dtype3$ where the corresponding bit-widths are such that $dtype1 \leq dtype2 \leq dtype3$. This implies that if $dtype1$ *takes 1 memory space* and $dtype2$ *takes 2 memory spaces*, then $dtype3$ *must take 2 or more memory spaces*. For this architecture, this is not always true. There exists a larger data-type ($dtype3$) which takes fewer memory spaces than a smaller data-type ($dtype2$). It is in a separate memory space and designed this way in order to meet the hardware timing requirements.

Many solutions were proposed. The first was to change the hardware. This was

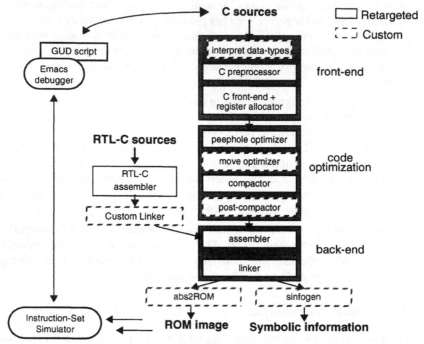

Figure 6.7 Full C Compiler suite and development environment

not possible because of timing restrictions. The second solution was to extend the memory handling of the compiler. A proposal was written for this solution which was to take several weeks to implement.

Instead of this solution, a third one was adopted. Although it took some time to conceptualize, it required only a small change to the compiler and took a relatively short time to implement. The details are somewhat complex, but the basic principle is as follows. The solution was to make the compiler interpret the larger data-type (*dt3*) as a smaller data-type (*dt2*) and vice-versa. Therefore, the smaller data-type would use more memory spaces than the larger data-type. The solution had only small side-effects. One was the requirement to make a small change to the compiler in interpreting constants so that constants used with the larger data-type were not chopped prematurely. A second concerned automatic variables placed on the stack, which would result in the waste of some memory spaces in some cases. This was partially resolved by providing some simple coding style rules.

Data-flow Optimization. As shown in Figure 6.6 the target architecture offers a considerable amount of parallelism. For example, the ALU and the ACU can work in parallel if they do not occupy the data bus at the same time. The instruction format provides orthogonal fields for parallel operations, so that compaction is rather straightforward (see Section 2.5 and Section 3.3.4). However, there are cases were data-flow optimizations must be performed in order to best exploit the available parallelism. The most important one is related to data moves within the ALU registers, which can be implemented either through the ALU, or through the data bus. The best choice is the one that allows another operation to be performed in parallel with the move, e.g. an ALU operation if the move is performed through the data bus, or any other operation occupying the data bus if the move is performed through the ALU. A custom *move optimizer* was implemented to improve the results of classic compaction. The utility keeps track of the operations that can be implemented in parallel with any move, while keeping track of the data-dependencies. It then selects the best move operation by evenly distributing the resource occupation, maximizing the potential parallelism, which is physically done later in the compaction phase.

Post Compaction. Although the processor has a 61 bit instruction word, the high amount of parallelism means that not all processor operations can be coded orthogonally. This means that certain sets of operations are chosen for parallel operation, while others must be sequential. In addition, the designer has chosen to implicitly encode operations into the instruction word by imposing restrictions on register usage, functional units, data-paths, etc.

This processor has a few encoding schemes which are beyond the capabilities of the compactor described in Section 3.3.4. To enhance the capabilities, a post-compaction phase was added which immediately follows compaction. This post compaction phase is built upon the peephole optimization approach. Rule are provided which search for sequences of assembly operations and replace these sequences with compacted sequences. Resources may be defined as part of a rule so that no data-dependencies are violated during compaction. We applied the post compactor to the specific encoding restrictions specific to this instruction-set. It performs well for all those regions of code where optimization is possible; however, the classic problem of coupling with other phases of code generation remained (e.g. register assignment is determined before compaction). While this cannot be avoided, we have found that the problem occurs rarely in practice.

ROM Generation and Custom Linker. Customizing the ROM contents for both the program and data memories was done to interface to the hardware environment. This was a straight-forward task of format conversion and was anticipated from the beginning of the project.

What was not anticipated was the development of a custom linker to integrate the macro assembly code (RTL-C) with the C application code. As explained earlier, the hardware refinement was done by means of writing time-critical portions of the code on a low level (RTL-C). This historical code investment is tapped only by integrating it with the application code written in C.

The linking strategy which was developed is shown in Figure 6.7. The binary code produced by the RTL-C compiler is treated as absolute data in a specific location. The assembler passes this block along with the code produced by the retargetable compiler to the linker. Since the code produced by the retargetable compiler is relocatable, the linker is able to find an absolute position other than the position of the RTL-C code.

Bit-True Library Development. Our methodology includes a path to compilation on the workstation for validation of the compiler and simulator, as described in Section 4.3. For algorithm verification, this is also an important path since compilation and execution on the host is typically an order of magnitude faster than execution on the instruction-set simulator.

For compilation on the workstation, the behavior must match the bit-true behavior of the processor. This implies the provision of a C library for the workstation which contains functions for each of the built-in functions defined for the retargetable compiler. For this architecture, the most important of these are the multiply-accumulate functions which have different behaviors depending on the data format being used. In addition, some functions perform automatic rounding and limiting operations.

Source-Level Debugging. The instruction-set bit-true model, developed to simulate the processor architecture, was implemented with a standard interpretative interface. It contains a set of interactive textual commands to run simulations, watch registers and memory contents, load data-files, etc.

Although it was not a priority at the departure of this project, a debugging interface for running the instruction-set simulator was desired. However, as the tools were maturing, this interface was re-evaluated as an essential part of the environment. It was the important piece that allowed the verification of correct operation of both the compiler and the instruction-set simulator.

The debugging interface was developed as an extension to the popular editor *Emacs*. It runs both under GNU Emacs (available from the Free Software Foundation [116]) and XEmacs (available from [122]). It is based on the GUD (Grand Unified Debugger) library which comes with Emacs. GUD also has similar interfaces for other debuggers (e.g. gdb, sdb, dbx, xdb, perldb). The interface is capable of setting break-points, cycle-stepping, C line-stepping, watching/printing registers, and printing global variables. The interface also includes an automatic retrieval of the

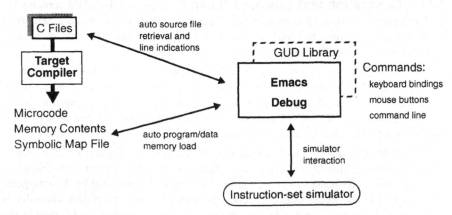

Figure 6.8 Debugging interface for the MMDSP C compiler

appropriate source file with automatic C line indications as shown in Figure 6.8.

To our surprise, only a minimum of source-level debugging information is needed in the symbolic map file to have a usable system (generated by *sinfogen* in Figure 6.7). The first version of the debugger contained only source line number information. This allowed running of a simulation by stepping through the source code as well as to set and run to break points, ensuring the correctness of the control-flow which was generated by the compiler. The second version of the debugger also contains global variable information.

Lessons in HW/SW Co-development. Throughout the development of this embedded system, a high interaction between the hardware and software teams took place. In addition to the high educational value of this concurrent design exercise, one main conclusion can be stated. Each side of the development has its set of complicated constraints. For problems on one side of the coin, the only way to reach a change on the other side is to push a little until the other side either finds a second way within his constraints, or pushes back.

This scenario was indicative in finding a solution to the memory and data-type problem described earlier. From the software side, the simplest solution would have been to change the hardware; however, from the hardware side, the simplest solution would have been to change the software. A formal negotiation following a study of the difficulties on each side was needed to resolve the problem, which by chance fell on the software side.

A similar issue arose which involved the operation of the program stack pointer. The original hardware operation of the pointer caused some very inefficient operation of function calls and returns on the compiler side. This would have resulted in slow operation and a waste of either program or data memory. Again, the easiest solution was to modify the operation of the hardware. In this case, this was a simple change in the hardware and was immediately carried out.

There are no straight-forward answers in the process of concurrent hardware/ software co-design. The process is an on-going challenge of staying within the constraints of both sides. The important aspect is a high-level of communication. And, at the very least, the interaction between the hardware and software teams allows solidification of the instruction-set specification, which serves as the formal contract between the two teams.

6.3.3 *Experimental results*

Compiler Validation. The compiler validation strategy that was used is described in Section 4.3. A set of validation tests was assembled in various categories which are summarized in Table 6.6. The first categorization breaks up the tests into two large categories: tests which can be used by a broad range of architectures, and others specifically for this architecture. The second categorization separates unit tests and full algorithmic type of tests.

Over 12000 lines of C code were run through the validation system, covering all

Table 6.6 Compiler validation tests for the TCEC MMDSP

Type of Test	Category	Operations, Functions	Number of C lines
Generic ANSI C	Unit Tests	bit-op, arith, relation, control, stack, ...	8742
	Integration Tests	bsearch, bubble, btree, gcd, wordcount, malloc, charcount, initptr	2842
Architecture Specific	Low/Medium Level Unit Tests	hardware loops, built-in functions, register sets, special registers	919
	Application Example	FFT	381
		Total	**12884**

the functionality of the processor that was expected to be used. This validates the stream of the firmware development environment including the retargetable compiler, instruction-set simulator, and bit-true library.

Table 6.7 Code size results of FlexCC for the TCEC MMDSP

Example	Number of Instructions Hand RTL-C	Number of Instructions FlexCC	Percentage Overhead
depack	80	High-Level (1) Source: 101	+26%
		Mid-Level (2) Source: 84	+0.5%
		Mid-Low Level (2-3) Source: 79	-0.1%
FFT	235	Mid-Level (2) Source: 261	+11%
		Mid-Low Level (2-3) Source: 228	-3%

Compiler Results. In successfully retargeting the compiler to this processor, the requirements set out at the beginning of the project were met. The compiler supports various levels of coding for different types of algorithms. We were able to evaluate the effects of these coding levels on two examples which were manually coded before the availability of the retargetable compiler. These results are shown in Table 6.7, which shows that for a high-level coding style a code size overhead of 26% is obtained. For a mid-level coding style, a code size overhead between 0.5% and 11% is obtained. The mid-low level coding style matches the code size of the manual code. While level 1 is the ideal level in the interest of clarity and portability, we have found that a mixture with levels 2 and 3 are necessary in time critical portions of the algorithms. For portions of the code which are not time critical, level 1 provides adequate code quality. It is interesting to note that level 4 (assembly-level) was never necessary, although it is a feature provided by the compiler.

Recap of Human Effort. Table 6.8 shows a breakdown of the effort spent on the various activities in the development of the compiler environment. The strong message from this breakdown is that roughly 30% of the effort was spent on validation of the compiler. This is an essential part of the design flow.

Table 6.8 Distribution of human effort by activity

Activity	Effort in Person-Months
Compiler Suite Retargeting	3.5
Custom Compiler Development	1.4
Compiler Validation	2.5
Support / Integration / Porting / Documentation	0.8
Total	**8.2**

Summary. The MMDSP firmware development environment includes a retargetable compiler, an instruction-set model, a source-level debugger, a validation strategy, and interfaces into the hardware environment.

The key lessons learned in this project are as follows:

1. Full environment: Although the compiler is the enabling technology, other tools are important to support the entire design activity. An instruction-set simulator and interfaces into the hardware design environment are critical parts. As well, the value of a source-level debugger cannot be underestimated.

2. Validation: A thorough validation test suite is mandatory, independent of the compiler approach. This constitutes nearly one third of the development effort, which includes the development of a bit-true library. If the application algorithms are available, these are of course the best validation benchmarks.

3. Compiler provision for low-level coding: Our experience shows that a code size overhead of about 30% is common for a high-level coding style. Our lesson was that the effort put into developing optimizations is a secondary priority after the requirement of the compiler to handle low-level coding styles. The designer must have control over the compiler so that he/she can meet his/her timing constraints when the compiler results are not satisfactory.

4. Concurrent design: Hardware and software should be developed concurrently in order to objectively evaluate the constraints on each. Concurrent development between hardware and software teams is always profitable.

5. Techniques which allows higher levels of coding are needed: Although point 3 is the industrial reality, compiler research should continue to find techniques which free the hardware designers from the software constraints.

6.4 Moving to higher coding levels

A behavioral level of C for embedded processors can only be supported by advanced compiler techniques which perform transformations based on the constraints of an architecture. For an effective transformation, the key elements are an architectural model and an explicit intermediate representation of the source algorithm. This next section describes experimental results of the address generation transformation for DSPs presented in Chapter 5.

6.4.1 DSP address calculation: experimental results

For an early version of the processor architecture described in Section 6.3.1, experiments were run using the ArrSyn array transformation approach. The specification and internal representation for that architecture is shown in Figure 6.9.

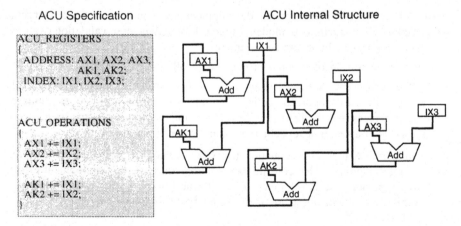

Figure 6.9 ArrSyn ACU specification for TCEC evaluation architecture

Working with a target compiler similar to the one described in Section 6.3.2, we compiled examples containing array references (1. High Level) with and without the prototype ArrSyn utility. These examples include various DSP functions, some specific to MPEG audio, others for standard DSP tasks such as interpolation and noise addition. Table 6.9 shows code size results, while Table 6.10 shows performance results. The values for Table 6.10 were calculated by assuming one cycle per instruction multiplied by the number of times a basic block was executed. This provides a first estimate of the performance, not taking into account conditional paths.

Table 6.9 Code size results of C compiler augmented by ArrSyn

Example	Number of Instructions C compiler	Number of Instructions ArrSyn + C compiler	% Improvement in Code Size
simple_loop	31	21	32%
median	83	56	33%
interpolate	72	49	32%
addnoise	59	48	19%
alloc	80	75	6%
Total	**325**	**249**	**23%**

Table 6.10 Performance results of C compiler augmented by ArrSyn

Example	Number of cycles C compiler	Number of cycles ArrSyn + C compiler	% Improvement in Time
simple_loop	103	69	33%
median	715	350	51%
interpolate	1017	499	61%
addnoise	1219	802	34%
alloc	10309	8526	17%
Average			**39%**

Table 6.9 and Table 6.10 show that a significant improvement in both the code size (23% reduction) and the performance (39% speed-up) resulting from the ArrSyn transformation. The explicit pointer addressing in the C code translates into a better utilization of the address calculation unit. Note that in these examples, the hardware do-loops which are available on the architecture have not as of yet been utilized. This

will lead to an additional improvement because of both the replacement of expensive branch instructions for hardware loop instructions as well as the removal of looping variables.

The conclusion that can be drawn from these results in combination with the results of Section 6.3.3 is that the calculation of addresses is the largest major factor which results in the inefficient compilation of C code in a high-level style for this type of architecture. We believe that this also applies to many DSP architectures with post-modify address calculation units. Furthermore, we have shown using the ArrSyn prototype that it is possible to improve the results of compilation with an automatic high-level transformation based on an architectural model.

6.5 Conclusion: compiler case studies in industry

This chapter has presented several case studies applying compiler techniques to a wide variety of embedded processor architectures in the fields of telecommunications and multimedia. Two compilation approaches have been used: model-based and rule-driven. While it is difficult to make a direct comparison of the two approaches as different architectures have many different needs, we can present some advantages and disadvantages of each approach. We attempt to keep our objectivity; therefore, comments will be restricted to the fundamental principles of each approach as there are many alien factors which contribute to the development of a project including software engineering and maintenance, company directions, and the constant changes to a development team.

It is possible to develop compiler algorithms well-adapted to architecture styles with a model-based approach. With a proper abstraction of the processor, a compiler can map source algorithms onto the architecture in a manner best suiting the data movement allowed in the structure. The ASIP developed at Nortel displayed a set of register characteristics canted toward special-purpose needs. As the CodeSyn compiler allows the description of register classes and structural abstraction of the architecture, it was possible to develop an instruction-set selection and register assignment approach driven toward special purposes. For a traditional approach to compilation, this task was shown to be very difficult as the homogeneous treatment of registers meant that registers in special roles could only be treated through reservation. The CodeSyn compilation results were shown to approach the quality of hand coded assembly programs, although further optimizations would be necessary to achieve the full performance of manual code.

In a setting where a compiler service is provided for a wide range of processor styles, the rule-driven approach has the advantage of covering the largest spectrum of machine types. It was shown to be possible to build compilers for architecture types varying from microcontrollers to VLIW DSPs. The approach has shown that the prototyping period is relatively short when setting up a compiler for a minimized architecture. For small architectures geared for very specific tasks such as the operators of the ST Integrated Video Telephone, retargeting time was shown to be on the order of one person month. However, the retargeting time has also been shown to be a strong function of the complexity of the architecture. Larger architectures such as the TCEC MMDSP supporting many data-types, multiple memories, and special operations needed a significant effort of approximately eight person months. In addition to the development time, roughly one third of the development time is needed for validation, independent of the compiler approach.

Furthermore, while the rule-driven, open programming based approach allows a development team to offer a compiler service to a wide range of processors, the support of architecture exploration is relatively low. Designers are less willing to modify an instruction-set specification for a compiler which has taken many months to develop. The user reconfiguration of a compiler for architectural modifications can only take place through a simplification of the instruction-set specification.

It is the author's belief that approaches founded upon architectural models describing the behavior of the processor are key to user retargetability. In the prototype work of the address calculation transformation ArrSyn, the strength of the approach was shown to be in the resource-based analysis. The mapping of the source algorithm to the target is best done using a behavioral abstraction of the mechanics of the architecture. The compilation algorithms are then based on the unique aspects of the processor operations.

On a final note, while model-based approaches show a great promise for a high degree of both architecture-based optimization and the ability for architecture exploration, we have learned from the experiences of using a rule-driven approach that it is important to maintain flexibility. The variation in existing processor design styles, architecture mechanics, special-purpose operations, and idiosyncracies can only be described as immeasurable. By consequence, a compiler developer must be *ready for anything*. Perhaps, a level of open programming which allows rules to be incorporated into a model-based compiler may be an effective route to the ultimate retargetable compiler. Just as embedded software allows late design changes for the system-on-a-chip, rules could allow late design changes for retargetable compilers!

Chapter 7: Tools for Instruction-Set Design and Redesign

While a retargetable compiler represents the principal implementation technology for the design of embedded systems, design exploration tools are also of great use for the development of a system. For example, a retargetable compiler does not provide many metrics to the designer of the processor, to measure how well the conception of the instruction-set has been done. Ideally, a maximum of feedback indicating the static and dynamic use of instruction codes is the type of information a designer would like to see during the design of the processor, and most certainly, as the processor evolves and is reused.

This chapter presents two new instruction-set design aids which allow a designer to analyze application code and conceptualize instruction level codings for a custom processor. Furthermore, a profiling tool is presented which permits algorithm exploration. The tools work together with a retargetable compiler methodology to allow the designer to explore design solutions of the instruction-set.

7.1 Tuning an instruction-set for different needs

For consumer electronics applications in multimedia and telecommunications, the most attractive feature of a programmable solution is the ability to track evolving standards with the flexibility provided by software. However, at what cost does a programmable solution have on the final product? Once in the product, the embedded system has a program which is *firm*, i.e. changed infrequently. Many questions can be posed: Are there available instructions of the processor that are never used? Is there hardware that could be taken out? Are there places where a new instruction could significantly improve the performance of the code?

The root of the issue is that an embedded processor's lifetime is extremely rich and long. In addition to the original design, many *flavors* of the architecture may be spun off for different reasons such as: modified applications, low cost versions, and in particular reuse for other products. This idea is shown pictorially in Figure 7.1.

When a product hits its market window, typically, the project does not stop. A designer may need to redesign a low-cost version of the product or simply need to evolve the product. The designer is then faced with the challenge of better refining the architecture with an understanding of how well it fits the existing application

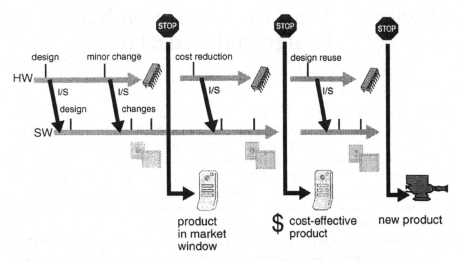

Figure 7.1 The rich lifetime of an embedded processor

code.

The needed tools in this area are those which furnish feedback of application code on a given architecture. This is shown in shaded blocks in the diagram of Figure 7.2, which is a subset of Figure 1.6 in Chapter 1. Ideally, the analysis tools should guide the designer by providing relevant statistics, and the possibility to make effective design changes based on those statistics.

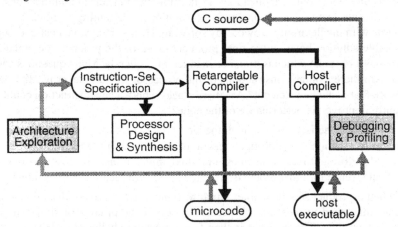

Figure 7.2 Tools for the design exploration of instruction-sets

7.2 Related work

Since the advent of instruction-set processors, quantitative approaches to architecture design have been proposed [43]. While these pioneering principles still hold for general computing systems, the constraints of embedded processor systems are putting an even larger importance on performance related architecture improvements.

The refinement of an instruction-set from a base superset of instructions have been proposed by Huang [48] and Holmer [45][46]. The main idea is to determine the most useful instructions from a base formula including parameters such as the execution profiles of compiled benchmarks. A model of the architecture data-path and the performance metrics allows the system to suggest good instructions to keep and to remove instructions of marginal benefit. While the statistical use is an important base principle, the approach has not yet been applied to embedded processors with architecture specialization and real-time constraints.

A similar approach has been proposed for the implementation of ASIP architectures in the PEAS system [4][50]. This approach attempts to minimize both software and hardware costs based on the compilation of source algorithms. A set of primitive operations allow compilation by the GNU gcc compiler, while a basic and extended set of operations may be included, based on the performance measures. For an objective area and power constraint, the performance is maximized by the system. However, optimization of the architecture below the primitive set of operations is not possible. The work of Breternitz and Shen [16] concentrate on a similar approach which uses a scalable Wide Instruction Word (WIW) architecture as an architecture template. A compiler and hardware allocator make the choices of functional units to include in the architecture.

The high expense of program memory for embedded processors has given rise to efforts of program width reduction such as instruction-word encoding (see Section 1.3.2). Some have approached the problem using a technique which reduces the width of the final program memory by exploiting redundancy [89]. Assuming all the application code is available, the entire program memory is divided into columns. Each instruction column which can possibly be generated from another instruction column is eliminated. This can be done through small hardware modifications such as exchanging multiplexer control lines and adding small pieces of logic. Nevertheless, this is a *last ditch* optimization that is not likely to allow the compiler to add any new software.

7.3 Overall flow

This section introduces two prototype design aids called *ReCode* and *ReBlock* shown in Figure 7.3. ReCode allows the exploration of the relationship between the instruction-set and the corresponding application code of custom embedded processors. After analyzing the instruction-set and code, the designer can then use the rich set of editing functions to adjust the instruction set to the application code. The designer can make gains by removing unused hardware, relieving bottlenecks in the hardware, removing unused instructions, and adding higher performance instructions. In conjunction, the utility checks the consistency of the instruction word while changes are being made. The tool works together with the instruction-set specification used by a retargetable compiler and can automatically regenerate the specification changes.

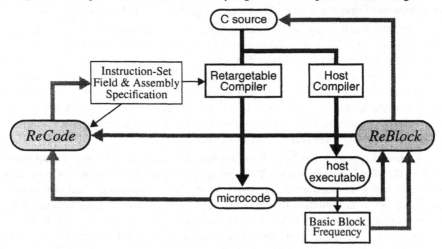

Figure 7.3 ReCode & ReBlock: tools for the analysis of application code.

In addition to static analysis of the application code, the tool can perform dynamic analysis with the link to the second tool ReBlock. ReBlock is a profiler requiring only the retargetable compiler and a host compiler for profile information. It automatically performs links between the microcode and the basic block executions on the host. Functions are available which estimate real-time performance based on the host execution. ReCode is able to work together with ReBlock to perform dynamic analysis of the instructions on either the basic block level or globally on a set of application files.

7.4 Analysis of application code

The user interface to ReCode is window-based and visually displays instruction-words in table form. The tool shares the instruction field and assembly specification file of the retargetable compiler. Alternatively, field and assembly encodings may be entered manually through the graphic interface.

Figure 7.4 shows the main window which displays the instruction word of the processor. Fields are shown as horizontal groupings of instruction bits with labels such as 'XXX' indicating the use of a set of bits within the instruction word. Each field has a set of assembly codes, which correspond to an encoding of the field. (Examples of assembly codes are shown in Figure 7.6.)

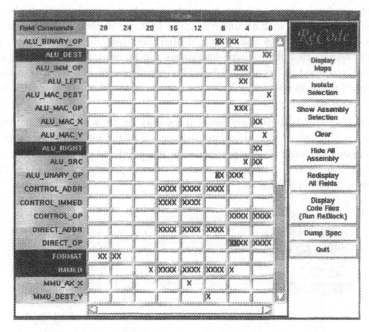

Figure 7.4 ReCode main window: display of instruction fields

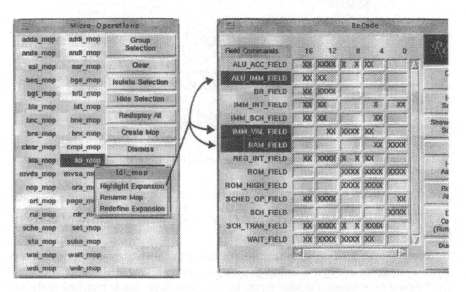

Figure 7.5 Expansion of micro-operations into instruction fields

The user has a set of commands which allow him to navigate through the field and assembly codings with an understanding of how the compiler uses the instruction-set. The compiler uses a set of micro-operations (*MOPs*) which expand into the different instruction fields. Micro-operations are higher-level compiler operations which are described in Section 3.3.1. MOPs are displayed as a set of buttons, each containing a function which highlights the expansion into the field entries of the main window as shown in Figure 7.5.

Following, fields and/or assembly lines may be shown or hidden from view which allows the developer to isolate one section of the instruction-set that can be worked upon. Analyses may be performed at this point and changes may be made to the coding using a set of editing functions. An example of field and assembly code isolation is shown in Figure 7.6.

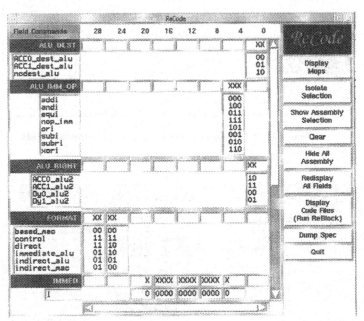

Figure 7.6 ReCode main window: isolation of fields and display of assembly

There are two main categories of functions:

Analysis Functions: These functions show the relationship between the instruction-set and the application software:

- Static Use of Assembly Codes: The frequency in which assembly codes are used may be generated in a distribution plot. Any combination of assembly codes may be used to compose the search pattern. This search pattern is matched to the machine code in vertical slices.

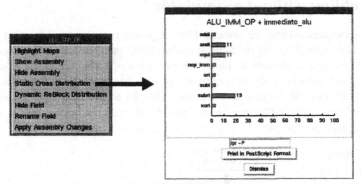

Figure 7.7 Field pull-down menu and static distribution plot

- Dynamic Use of Assembly Codes: Linking to the ReBlock profiler, dynamic use of assembly codes may be shown in a distribution plot. This can be done on selected regions in the code to resolve bottlenecks, or on all the application code to identify critical points in the hardware.

The distribution plots are activated by a pull-down menu on each of the fields in the main menu. Figure 7.7 shows an example of the pull-down menu and a distribution plot for the ALU_IMM_OP field combined with the immediate_alu assembly code of the instruction-set shown in Figure 7.6.

Static and dynamic analyses may also be done on resource activation fields, for example, busses and register files. For example, the distribution of data which moves from one register file to another over a data bus is information useful for the hardware designer. This allows the identification of both low resource use and the appearance of congestion spots.

Editing Functions: Working together with the analysis functions, the designer is able to use the editing functions to modify the instruction-set according to his/her requirements. These functions include:

- Consistency checking of assembly bit changes: The user is free to make changes which when applied will be directly reflected in the specification file. Verifications are made after the application of changes. Shortcut button touches are provided for autocoding ascending, descending and other fill patterns as well as the removal of unused assembly codes.

- Bit-field width reduction: Given a reduced number of assembly codes within a field, the width of the field may be reduced automatically.

- Regeneration of the instruction-set specification. Following an interactive session of ReCode, the instruction-set specification can be regenerated for the retargetable compiler. A repass of the retargetable compiler for all the application code is then possible. The resulting code can once again be analyzed by ReCode.

7.5   Profiling without a simulator

Basic block frequencies may be generated by the profile function (option -a) of many host compilers, including the publicly available GNU gcc [96]. This function adds extra code to a source file which counts the execution of basic blocks. After execution on the host, the frequency of occurrence of each basic block is written to a file. These basic blocks can be linked to the basic blocks of the retargetable compiler machine code through the C line numbers intended for debugging. Figure 7.8 shows this process with an example.

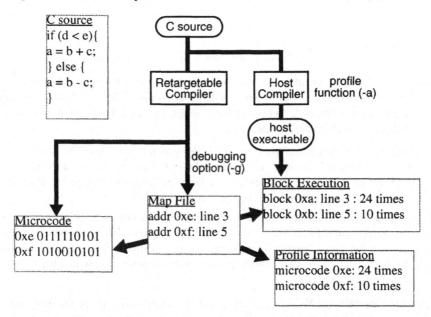

Figure 7.8 Generating profile information through basic block links.

Putting the pieces together allows the construction of a profiling browser: the ReBlock profiling interface is shown in Figure 7.9. The left highlighted column shows the correspondence to microcode addresses of the target compiler. The user can see the number of microcode instructions which correspond to each part of the C source. The second highlighted column shows the profile executions for each basic block of the host execution. These two columns may be shown or hidden from view allowing the interface also to be used as a simple editor.

As a consequence of the available information, this methodology can be used to effectively estimate real-time performance. Counting the number of assembly lines within each basic block and multiplying by the frequency gives the number of instruction cycles to be executed. Subsequently dividing by the instruction clock rate (either estimated or real) can give an excellent first order approximation of the final speed of the code. This real-time performance information can be used to redesign

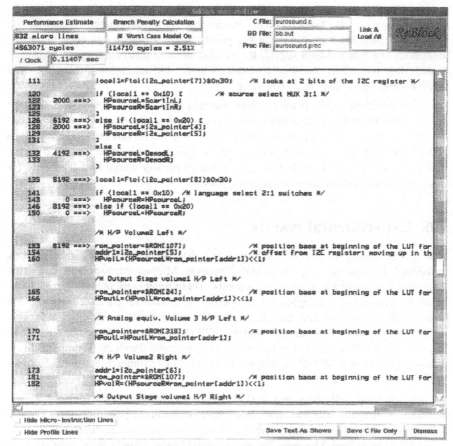

Figure 7.9 User-interface for the ReBlock profiler

either the processor or the algorithm. Note that this methodology generates an estimate of real-time execution without the existence of an instruction-set simulator. This
is important since the approach can be used in the design stage before a simulator is
implemented. Furthermore, long simulations would favor this approach as an instruction-set simulator will run, at best, an order of magnitude slower than execution on
the host processor.

The method assumes that instructions have a fixed cycle count. For processors
with instructions of varying cycle count (e.g. CISC), a more precise correspondence
to the instruction-set is necessary. This would be possible with a precise link to the
instruction-set specification (e.g. via ReCode), but has not yet been implemented.

The ReBlock approach produces an estimate of real-time performance which
does not take into account dynamic effects. For example, should the processor contain an instruction or data cache, cache-misses would incur delays. Another more
common feature in embedded processors is a pipelined execution controller. Jumps

or branches of the program incur an instruction-cycle penalty if the jump is taken. In ReBlock, a worst-case branch penalty model has been incorporated. Based on the basic block profiling executions, all the points where a branch has been taken are identified. The user can then specify the number of extra instruction cycles that each branch may take. This worst case model is useful for determining the upper and lower bounds on the total branch penalty.

This worst-case branch model does not take into account delay slots that the compiler has filled for delayed branches and returns that may exist in the instruction-set. A more precise branch penalty model would be possible; however, this requires a semantic knowledge of the branch instructions. This again could be furnished by the instruction-set specification. In this case, a branch penalty would be added only for those instructions which trigger a pipeline stall.

7.6 Experimental results

Preliminary testing of ReCode and ReBlock has been done using five existing embedded processors, four from SGS-Thomson Microelectronics and one from Thomson Consumer Electronic Components. These processors include three blocks of the Integrated Video Telephone developed in the Central Research & Development Group at SGS-Thomson and described in Section 6.2.1: the BSP (Bit-Stream Processor), the MSQ (MicroSeQuencer), and the VIP (VLIW Image Processor); the DAP (Digital Audio Processor) developed in the Dedicated Products Group of SGS-Thomson; and the MMDSP developed at Thomson Consumer Electronics Components described in Section 6.3.1.

This section will show just a skeleton of the possible uses of ReCode and ReBlock, illustrating with examples from the aforementioned architectures. The following does not provide suggestions for changes to these architectures, instruction-sets, or the algorithms. The compiler environments and the application code are all in some intermediate phase at the time of this writing. The examples are merely illustrations of the use of the ReCode and ReBlock tool set so that designers can perform similar analyses.

7.6.1 Operation instruction code usage

The ReCode utility may be used to determine how well instruction codes for a given machine are matched to the application code. We present examples using the MSQ architecture of the ST IVT and a number of H.261 algorithm benchmarks covering 1825 lines of assembly code: grabber, idct_out, motion, scheduler, polling_loop, and host_if. Figure 7.10 shows examples from two of the main functional units of the architecture, the ALU (Arithmetic and Logical Unit) and the branch commands of the sequencer unit. For the ALU, Figure 7.10 a) shows that both the ADDA and ANDA operations are entirely unused in all of the application code. Because immediate operations ADDI and ANDI are used in the software means that

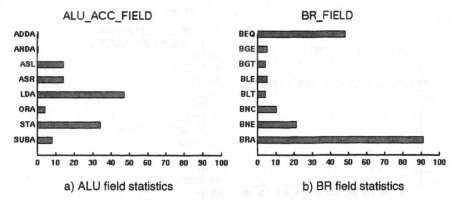

a) ALU field statistics   b) BR field statistics

Figure 7.10 Distribution usage of assembly codes for the ST IVT MSQ core.

the hardware operations cannot be removed from the functional unit; however, the designer can make the consideration of removing these assembly codes for use elsewhere or for eventual reduction of the instruction width. The second aspect of note is that the LDA (load accumulator) and STA (store accumulator) codes are used more than twice the amount of times than any other operation. This high memory access for the accumulator may be a reason for considering an increase in the number of accumulators.

For the BR field (Figure 7.10 b), it is interesting to note that the BRA (unconditional branch) code is used nearly twice as much as any other branch code. This could indicate to the compiler designer that branch or block reorganization optimizations may be a wise area for investment. On the other hand, it could also indicate to the hardware designer that a delayed branching mechanism may be of use.

For the MMDSP processor, we conducted similar experiments on a small set of available examples (574 assembly code lines). For the DCU (data calculation unit), we found that only 22 of the available 64 operations were used. This means that a reduction of the field width by one (from $2^6=64$ to $2^5=32$) would limit the number of operations supported. However, it would still leave 10 opcodes for operations in the application code yet to be written. In further investigations, we performed an analysis of immediate operations with the DCU. These operations take one of 2 operands directly from the instruction-word. The analysis showed that although the unit is capable of routing the immediate value from either the left or right side of the DCU, it only ever uses the left side for immediate values. Removal of the immediate to the right of the DCU would give a one bit savings, with the additional savings of hardware. Using only the left immediate of the DCU is a heterogeneous characteristic of the compiler which illustrates an area of possible instruction-word savings.

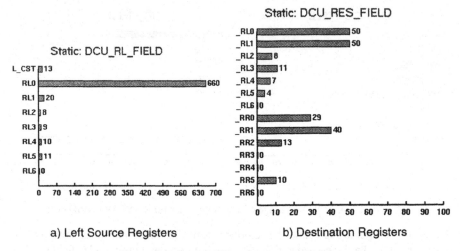

a) Left Source Registers   b) Destination Registers

Figure 7.11 Static DCU register usage distribution

7.6.2 *Data occupation and movement*

ReCode can be used to statistically measure the characteristics of a compiler-machine relationship. For example, the register file usage indicates how data is routed through the machine according to the application code. Results indicate two things:

• How the compiler uses the available registers.

• How restrictive the instruction-set of the machine is on register usage.

Analyzing the MMDSP architecture, we show a part of the register file usage of the DCU (Data Calculation Unit) in Figure 7.11. The left plot (Figure 7.11 a) shows the use of registers in the left source register file and the right plot (Figure 7.11 b) indicates the register usage of the output of the DCU. For the left source, removing the number of times the DCU was not used (NOP) from the default case (RL0) still leaves RL0 used roughly 100 times, which is 5 times more than any of the other registers (RL6 is not used at all). This is partly because the processor has special-purpose uses of RL0, and partly the choice of the register allocation of the compiler. For the destination registers, the distribution is also leaning toward the first registers in the set; however, it has a much more distributed use of the available registers than the DCU input. For this set of code, the input of the DCU requires much less freedom of register choice than the output. This is an interesting result for both the compiler developer and the hardware designer who may want to measure the trade-off of constraining more registers for other special purposes.

In further analyses of this architecture, we turned our attention to general data movement. This processor allows register to register movement by means of a bus. This includes registers of the DCU, ACU, sequencer unit and processor interface.

The bus instructions use $2^6 + 2^6 = 64 + 64 =>$ 12 bits to specify the transfer, meaning 64 registers may move to any other 64 registers. Again, doing similar statistics on the application code, we determined that only 24 distinct registers are used as sources and 30 distinct registers are used as destinations. This means that by choosing a good subset of transfers, this field can possibly be reduced to $2^5 + 2^5 = 32 + 32 => 10$ bits. Of course, this must be done with care, always allowing the needed data movement used by the compiler. Other (possibly slower) data paths exist in the architecture and can be easily matched to the less frequently used paths in the application code.

7.6.3 Algorithm and instruction-set profiling

Using the ReBlock profiling approach described in Section 7.5, a number of performance calculations can be made on source algorithms very early in the architecture development. For example, Figure 7.12 shows a report summary generated by ReBlock after an interactive use of the profiler for the Eurosound application running on the SGS-Thomson Digital Audio Processor (DAP). The top part of the report

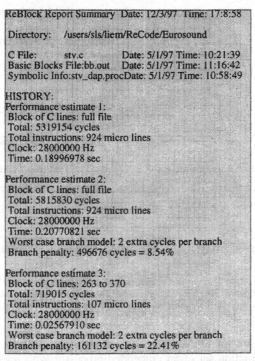

Figure 7.12 Report generated by ReBlock profiler on Eurosound application

shows the files being analyzed and their respective timestamps, while the bottom part of the report contains information on which performance analyses have been run. The first estimate shows the full C file being analyzed and a calculation of the real-time performance based on the user-specified clock. The second estimate shows the same analysis, this time with the addition of the worst-case branch penalty (2 cycles per branch) which would account for nearly 9% of the total application performance. The third estimate is focused on a region of C code (lines: 263 to 370) where the worst-case branch penalty would account for over 22% of the performance of this region. The user is free to make estimates on any parts of the code, playing *what if* games with the parameters such as the clock frequency and the number of cycles for each branch penalty.

In addition to speed calculations, the link from the instruction-set analysis tool, ReCode, to the profiler, ReBlock, allows the statistical use of assembly codes to be augmented by their dynamic use. For example, in Figure 7.13, we show both the static versus dynamic use of ALU immediate operations in the same ST DAP architecture for the Eurosound application. Notice that although the equi instruction appears the most frequently in the microcode, it is executed many thousands of times less than the other instructions andi and subri. These may be important measurements for improving the performance of the instruction-set architecture.

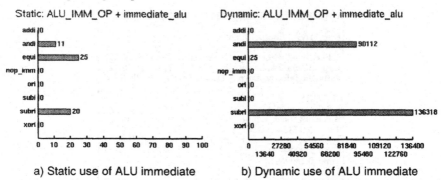

a) Static use of ALU immediate   b) Dynamic use of ALU immediate

Figure 7.13 Example of static versus dynamic instruction statistics for DAP architecture

7.7 Chapter summary: instruction-set exploration tools

This chapter has presented a toolset for instruction-set design of embedded processors. These design aids serve to support the naturally long lifetime of an embedded processor architecture as it evolves from an initial product to a cost reduction and is possibly reused for new products.

The ReCode instruction analysis tool allows a designer to perform measurements of the instruction-set of an embedded processor with regard to the existing application code. The analysis functions include the ability to show distribution of assembly codes, both in static and dynamic form. Using this information, a set of editing functions are provided to the designer such that the instruction encoding may be modified and the specification for the retargetable compiler be automatically regenerated. At this point, a compilation may be re-run for the modified instruction-set.

The ReBlock profiling tool allows performance analyses of microcode, linking the results of the retargetable compiler with the execution on a host. Thus, the dynamic information is generated without the use of a dedicated simulator, which means the analysis can be done before the simulator is implemented. Furthermore, the execution run-time would normally be an order of magnitude faster than interpretive instruction-set simulation. The ReBlock profiling information is also used to support the dynamic analysis functions of ReCode.

Experiments have been done with a set of existing processors. Example analyses were presented showing results which are both useful for the embedded system designer and programmer. These include aspects of functional unit usage, register and data-bus occupation, and performance-oriented calculations including worst-case branch penalties and dynamic use of instructions. The information generated by these analyses may also be useful to the compiler developer wishing to improve his/her algorithms.

Many avenues for future work lie ahead. The natural first step is to conduct more experiments with new architectures. The tools are currently being explored by two design groups at SGS-Thomson Microelectronics. With the number of possible new features, additions will be made with priority to designer requests. The set of current requests include the addition of an application programming interface (API) to ReCode, more possibilities on the reorganization of statistics, and the ability to perform critical path analysis with ReBlock.

As well as these incremental improvements to the toolset, the approach has opened some avenues for larger projects in ASIP design. One of the most tedious tasks in the implementation of the processor hardware is the HDL description of the instruction-word decoder. This is especially true if there are changes to the instruction-set during the refinement of the architecture and throughout its lifetime. A link from the instruction-set to the functional unit behavior of the architecture could possibly be a way to generate this description.

A second project is in the area of low power coding. Given that both the instruction-level coding and the execution profiles of the application are available with the ReCode/ReBlock toolset, it would be possible to implement an instruction-set recoding algorithm which minimizes the power utilization of embedded software on a processor. This could even be brought one step further taking into account the power draw of functional units, should the previously mentioned project be in place. The largely untouched domain of ASIP design is open for new ideas in research and development.

Chapter 8: Conclusion

8.1 Summary and contributions

This book has presented a wide range of aspects regarding the application of compiler methodologies to embedded processor systems. Retargetable compilation techniques have been shown to be an important design technology for the constraints of today's embedded architectures. To begin, an evolution of embedded processor architectures was presented, showing the application pull on the characteristics of instruction-set architectures. It is recognized that real-time constraints put significant demands on the performance of embedded processors such as DSPs. This is manifested in RISC, pipelined, and VLIW principles for architectures with specialized functional units and register structures.

Furthermore, memory real estate is being acknowledged as one of the most important resources, especially for the system-on-a-chip. This has given rise to instruction-word encoding schemes for DSP architectures which improve the program memory utilization. For DSP as well as MCU architectures, instruction-word minimization is also manifested in unusual program memory management like program paging.

These new embedded architectures have brought about new challenges for modern compilation techniques. An overview of traditional and emerging compiler techniques have been presented with respect to these embedded processors. A wealth of appropriate techniques have been discussed; however, a number of areas need to be further researched and developed.

Following, two practical retargetable compiler systems were presented. The CodeSyn compiler developed at Bell-Northern Research/Nortel is based upon a model of the architecture including the behavior of instructions, the structural connectivity, and a set of resource classes. The advantage of the approach is the ability to automatically retarget the system by changes to the model. The algorithms for compilation are all centralized by the architecture model. In particular, enhanced pattern matching and selection algorithms as well as register allocation and assignment algorithms for special-purpose registers were shown possible. A disadvantage of the approach is that processor targets must lie within the boundaries of the architecture model and any peculiarities of a new target must be treated by updating the algorithms.

The FlexCC compiler used by SGS-Thomson Microelectronics is based on a

141

rule-driven approach. The compiler has many traditional compilation steps that have been reorganized in an open-programming environment. Each step allows the execution of parameterized rules which manipulate the transformation of code to the target. The advantage of the approach is the flexibility for a very wide set of architectures. Standard rules can be put in place for most cases, while the developer can concentrate on processor idiosyncracies. Results are heavily dependent on the development effort put into optimization and retargeting time is strongly dependent on the complexity of the architecture. A disadvantage of the approach is that it requires an expert's development time, which makes architecture exploration difficult.

Next, practical issues for firmware development environments were discussed. These include language support, coding styles, compiler validation strategies, and source-level debugging. All of these are important considerations for projects in industrial environments.

In Chapter 5, an approach was proposed targeting the address calculation units of DSP architectures. As memory access is a particularly important performance consideration for signal processing, this type of transformation is critical for an effective compiler. The approach introduces a flexible architectural model and a compiler transformation for address generation. The system can be easily integrated into any target compiler to improve performance results. Furthermore, the simplicity of the specification allows a designer to do architecture design space exploration.

Chapter 6 summarizes the application of many of the principles and techniques presented in Chapters 2 to 5 to a set of processors used in industry. The processors include a telecommunications ASIP developed at Nortel, a set of operators for the Integrated Video Telephone at SGS-Thomson Microelectronics, and the MMDSP developed at Thomson Consumer Electronics Components. The main conclusion is that model-based transformations show a promising avenue for high performance compilation to embedded processors. However, it remains important to have a significant level of open-programming, the strength of the rule-driven approach. A number of lessons came out of the experiences including the need for full firmware development environments, the need to reserve approximately one third of the development time for validating a compiler environment, the need for the low-level coding style support in a retargetable compiler, and the need for high communication of hardware and software development teams.

Finally, Chapter 7 introduces a set of application and architecture exploration tools which provide feedback and analyses to the designer of an embedded system. These tools allow the statistical analysis of instructions in application code in both static and dynamic modes. Furthermore an approach for performance profiling was proposed which does not require the full development of an instruction-set simulator. The methods fit into a retargetable compiler methodology providing a means for exploration to the embedded system designer.

For the area of embedded system design, the projects and developments described in the text have made its main contributions in three principal areas:

- Practical methodologies and experiences with retargetable compilation in industrial applications of embedded systems.

- A model-based transformation which performs efficient address generation for post-modify based address calculation units.

- A set of exploration tools which permit a designer to refine either an architecture or algorithm within its application domain.

8.2 What's ahead?

In comparison to compilation for general purpose processors, the *savoir-faire* in retargetable compilation for embedded processors is currently in its infancy. The industry is slowly making steps which are improving the situation; however, the advances can only be described as a *crawl* in comparison to the advances in the technologies of embedded processor architectures. Today's popular solution to the poor state of embedded software development tools is simply to hire more engineers who inevitably code on the assembly level. It is surprising to see the large number of job opportunities today for engineers with embedded processor assembly experience. As more and more assembly lines of code are written, companies become locked to old architectures, systems become more sophisticated, and the embedded software crisis mounts. A revolution of compilation techniques is needed to revise the electronic industry's traditional view of embedded software.

On the other hand, the designers of embedded processors are the first to see the importance of software tools for both the design *with* and the design *of* embedded architectures. Even they are beginning to put together their own embedded software tools from scratch. The need for tools for embedded processor systems coupled with the emerging popularity of hardware-software co-design leaves us little doubt that the revolution in design technology for embedded processors will indeed arrive.

How this technological revolution will present itself is an open question. The research community appears to be resting its hopes on the path toward completely retargetable compilers, where a simple specification of the processor is enough for the tool to reconfigure all its transformations. I remember some discussions back at Nortel wondering how far we could take this concept. We imagined this ultimate tool that could take anything a designer could stuff into it: behavioral descriptions, RTL descriptions, netlists, even the chip itself plugged into a socket and - *wham, zap, presto* - out comes a brand new optimized compiler for the hardware!

Having seen the wide variety of architectures that exist today, it is not likely the case that this extreme goal will be achieved. If we were to compare retargetable compilation to the growing area of behavioral synthesis, while they are analogous on the level of techniques they perform their respective tasks based on two very different premises. The behavioral synthesis process *constructs* an architecture from a set of components with a nearly unrestricted set of resources. The only constraints are those imposed by the designer, such as a time deadline or a power budget. On the other hand, a retargetable compiler does not have the freedom to construct. It is obliged to *conform* to the fixed architecture. The transformation algorithms do not merely have constraints as objectives; these constraints *are* the specification! A programmable processor could function on any wide number of principles that some ingenious (*or nutty!*) designer has dreamed up. The chances that a compiler specification can adequately describe any possible processor design style are slim.

While we can abandon the idea of a fully retargetable compiler, we can well imagine the possibility of restricted retargetability within architecture *flavors*. Once one category of processors is handled well by a compiler, it is foreseeable that flexibility be incorporated so that it can be parameterized; and therefore, the range of compiler retargetability would be restricted to a well-known set. This point makes Goossen's work [36] on the classification of instruction-set processors by their properties an important contribution.

Just as we identified the importance of customizing an architecture to the needs of an application, software tools which can easily match the processor customization through compiler parameterization will be an added competitive factor. By the way, that applies not only to instruction-set processors but to any programmable system design which includes hardware and software.

Bibliography

[1]   A. Aho, R. Sethi, J. Ullman, Compilers: Principles, Techniques and Tools, Addison-Wesley, Reading, Massachusetts, 1988.

[2]   A. Aho, S. Johnson, "Optimal code generation for expression trees", *Journal of the ACM*, Vol 13, no 3, July 1976., pp. 488-501.

[3]   A. Aho, M. Ganapathi, S. Tjiang, "Code generation using tree matching and dynamic programming", *ACM Transactions on Programming Languages and Systems* 11, 4, Oct. 1989, pp. 491-516.

[4]   A. Alomary, T. Nakata, Y. Honma, M. Imai, N. Hikichi, "An ASIP Instruction Set Optimization Algorithm with Functional Module Sharing Constraint, *Proc. of the International Conference on CAD*, 1993, pp. 526-532.

[5]   Analog Devices, "ADSP--2100 Family DSP Microcomputers", available at http://www.analog.com

[6]   G. Araujo et. al., "Challenges in Code Generation for Embedded Processors", in Code Generation for Embedded Processors, ed. by P. Marwedel, G. Goossens, Kluwer Academic Publishers, 1995.

[7]   G. Araujo, S. Malik, "Optimal Code Generation for Embedded Memory Non-Homogeneous Register Architectures", *International Symposium on System Synthesis,* Cannes, France, Sept. 1995, pp. 36-41.

[8]   R. Camposano, J. Wilberg, "Embedded System Design", *Design Automation for Embedded Systems, an International Journal,* Kluwer Academic Publishers, 1996, v. !, pp. 5-50.

[9]   D. Bacon, S. Graham, O. Sharp, "Compiler Transformations for High-Performance Computing", *ACM Computing Surveys,* Vol. 26, No. 4, December, 1994, pp. 345-420.

[10]   U. Banerjee, Loop Parallelization, Kluwer Academic Publishers, 1994, 171 pages.

[11]   S. Bashford et. al., "The MIMOLA Language Version 4.1", Technical Report, Lehrstuhl Informatik XII, University of Dortmund, Sept. 1994.

[12]   L. Bergher, X. Figari, F. Frederiksen, M. Froidevaux, J.M. Gentit, O. Queinnec, "MPEG Audio Decoder for Consumer Applications", *Proc. of the Custom Integrated Circuits Conference*, May 1995, Santa Clara, Ca.

[13]   Berkeley Design Technology Inc., DSP Design Tools and Methodologies, 1995, see http://www.bdti.com

[14]   E. Berrebi, P. Kission, S. Vernalde, S. DeTroch, J.C. Herluison, J. Fréhel, A. Jerraya, I. Bolsens, "Combined Control-flow Dominated and Data-flow Dominated High-level Synthesis", *Proc. of the Design Automation Conference*, June 1996, pp. 573-578.

[15]   J. Bier, "DSP Processors and Cores: The Options Multiply", *Integrated System Design Magazine*, June, 1995.

[16]   M. Breternitz, J. Shen, "Architecture Synthesis of High-Performance Application-Specific Processors", *Proc. of the Design Automation Conference*, 1990, pp. 542-548.

[17]   G. Chaitin et. al., "Register Allocation via Coloring", in *Computer Languages*, Pergamon Press Ltd., Vol. 6, Jan. 1981, pp. 47-57.

[18]   G. Chaitin, "Register Allocation & Spilling via Graph Coloring", *Proc. of the ACM Symposium on Compiler Construction*, SIGPLAN Notices, Vol. 17, no. 6, June 1982, pp. 98-105.

145

146

[19] The Corporate Software Integrator, "Lode DSP Engine: Preliminary Data Sheet", May 1995.

[20] J. Davidson, C. Fraser, "Code Selection through Object Code Optimization", *ACM Transactions on Programming Languages and Systems*, Vol. 6, No. 4, October 1984, pp. 505-526.

[21] F. Depuydt, W. Geurts, G. Goossens, H. DeMan, "Optimal Scheduling and Software Pipelining of Repetitive Signal Flow Graphs with Delay Line Optimization", *Proc. of the European Design and Test Conference*, 1994, pp. 490-494.

[22] DWARF Debugging Information Format, Unix International Programming Languages Special Interest Group, Parsipanny, NJ, Revision 2.0.0, July 27, 1993

[23] H. Emmelmann, P. Schroer, R. Landwehr, "BEG - a generator for efficient back ends", *ACM SIGPLAN Conference on Programming Language Design and Implementation*, Vol. 24, No. 7, July 1989, pp. 227-237.

[24] A. Fauth, "Beyond tool specific machine descriptions", in Code Generation for Embedded Processors, ed. by P. Marwedel, G. Goossens, Kluwer Academic Publishers, 1995.

[25] A. Fauth, A. Knoll, "Automated generation of DSP program development tool using a machine description formalism", *Proceedings of ICASSP*, 1993.

[26] A. Fauth, J. VanPraet, M. Freericks, "Describing Instruction Set Processors Using nML", *Proc. of the European Design and Test Conference*, 1995, pp. 503-507.

[27] H. Feuerhahn, "Data-Flow Driven Resource Allocation in a Retargetable Microcode Compiler", *Proc. of the 21st Workshop on Microprogramming and Microarchitecture*, 1988, pp. 105-107.

[28] C. Fischer, R. LeBlanc, Crafing a Compiler wih C, The Benjamin/Cummings Publishing Co., Redwood City, Ca, 1991.

[29] F. Franssen, F. Balasa, M. van Swaaij, F. Catthoor, H. DeMan, "Modeling Multi-Dimensional Data and Control flow", *IEEE Trans. on VLSI Systems*, Sept. 93, Vol. 1, pp. 319-327.

[30] C. Fraser, D. Hanson, T. Proebsting, "Engineering a Simple, Efficient Code Generator Generator", *ACM Letters on Programming Languages and Systems*, Vol. 1, No. 3, Sept. 1992, pp. 213-226.

[31] M. Freericks, "The nML Machine Description Formalism", Technical Report 1991/15, TU Berlin, Fachbereich Informatik, Berlin, 1991.

[32] M. Ganapathi, C.N. Fisher, and J.L. Hennessy, "Retargetable compiler code generation", *ACM Computing Surveys*, Vol. 14., 1982, pp. 573-593.

[33] N. Gehani, C: An Advanced Introduction, ANSI C Edition, Computer Science Press, Inc. New York, 1988.

[34] W. Geurts, et. al., "Design of DSP systems with Chess/Checkers", *2nd International Workshop on Code Generation for Embedded Processors*, Leuven, Belgium, March 18-20, 1996.

[35] R. Glanville, S. Graham, "A new method for compiler code generation", *Proc. of the 5th ACM Symposium on Principles of Programming Languages*, 1978.

[36] G. Goossens, J. VanPraet, D. Lanneer, W. Geurts, A. Kifli, C. Liem, P. Paulin, "Embedded Software in Real-Time Signal Processing Systems: Design Technologies", *Proceedings of the IEEE, special issue on Hardware/Software Co-Design*, 1997.

[37] G. Goossens, J. Vandewalle, H. DeMan, "Loop optimization in register-transfer scheduling for DSP-systems", *Proc. of the Design Automation Conference*, 1989, pp. 826-831.

[38] R.P. Gurd, "Experience Developing Microcode Using a High-Level Language", *Proc. of the 16th Annual Microprogramming Workshop*, Oct 1983, pp. 179-184.

[39] M. Harrand et al., "A Single Chip Videophone Encoder/Decoder", *Proc. of the IEEE International Solid-State Circuits Conference*, Feb. 1995, pp. 292-293

[40] R. Hartmann, "Combined Scheduling and Data Routing for Programmable ASIC Systems", *Proc. of the European Design Automation Conference*, 1992, pp. 486-490.

[41]   L. Hendren, G. Gao, E. Altman, C. Mukerji, "A Register Allocation Framework Based on Hierarchical Cyclic Interval Graphs", *Proc. of the International Conference on Compiler Construction*, 1992.

[42]   J. Henkel, R. Ernst, U. Holtmann, T. Benner, "Adaptation of Partitioning and High-Level Synthesis in Hardware/Software Co-Synthesis", *Proc. of the International Conference on CAD*, 1994, pp. 96-100.

[43]   J. Hennesey, D. Patterson, Computer Architecture: A Quantitative Approach, Morgan Kaufmann Publishers, San Mateo, CA, 1990.

[44]   P. Hilfinger, J. Rabaey, "DSP specification using the SILAGE Language", in Anatomy of a Silicon Compiler, ed. by R.W. Brodersen, Kluwer Academic Publishers, 1992.

[45]   B. Holmer, "A Tool for Processor Instruction-Set Design", *Proc. of the European Design Automation Conference*, 1994, pp. 150-155.

[46]   B. Holmer, A. Despain, "Viewing Instruction Set Design as an Optimization Problem", Proc. *of the 24th Annual International Symposium on Microarchitecture*, Nov. 1991, pp. 153-162.

[47]   U. Holtmann, R. Ernst, "Combining MBP-Speculative Computation and Loop Pipelining in High-Level Synthesis", *Proc. of the European Design & Test Conference*, 1995, pp. 550-555.

[48]   I. Huang, A. Despain, "Synthesis of Application Specific Instruction Sets", *IEEE Transactions on Computer-Aided Design of Integrated Circuits and Systems*, v. 14, no. 6, June 1995, pp. 663-675.

[49]   M. Ikeda et al., "A Hardware/Software Concurrent Design for a Real-Time SP@ML MPEG2 Video-Encoder Chip Set", *Proc. of the European Design & Test Conference*, 1996, pp. 320-326.

[50]   M. Imai, A. Almary, J. Sato, N. Hikichi, "An Integer Programming Approach to Instruction Implementation Method Selection Problem", *Proc. of the European Design Automation Conference*, 1992, pp. 106-111.

[51]   T. Ben Ismail, K. O'Brien, A. Jerraya, "Interactive System-level Partitioning with PARTIF", *Proc. of the European Design & Test Conference*, 1994.

[52]   S. Joshi, D. Dhamdhere, " A composite hoisting-strength reduction transformation for global program optimization", *International Journal of Computer Math*, part 1., Vol. 11, No 1., 1982, pp. 21-41, part 2, Vol. 11, No. 2.

[53]   J. Knoop, O. Ruthing, B. Steffen, "Lazy Code Motion", *ACM SIGPLAN Conference on Programming Language Design and Implementation*, 1992, pp. 224-234.

[54]   F. Kurdahi, A. Parker, "REAL: A Program for REgister ALlocation", *Proc. of the 24th Design Automation Conference*, 1987, pp. 210-215.

[55]   Lam M., "Software pipelining: An effective scheduling technique for VLIW machines", *ACM SIGPLAN Conference on Programming Language Design and Implementation*, Vol. 23, No. 7, July 1988, pp. 318-328.

[56]   D. Lanneer, M. Cornero, G. Goossens, H. DeMan, "Data-routing : a paradigm for efficient data-path synthesis and code generation", *Proc. of the 7th Int. Symposium on High-Level Synthesis*, May 1994, pp. 17-22.

[57]   D. Lanneer, et. al., "Chess: Retargetable code generation for embedded DSP processors", in Code Generation for Embedded Processors, ed. by P. Marwedel, G. Goossens, Kluwer Academic Publishers, 1995.

[58]   J.R. Larus, "Efficient Program Tracing", *IEEE Computer*, May 1993, pp. 52-61.

[59]   R. Leupers, P. Marwedel, "Time-constrained Code Compaction for DSPs", *International Symposium on System Synthesis*, Cannes, France, Sept. 1995, pp. 54-58.

[60]   R. Leupers, P. Marwedel, "Algorithms for Address Assignment in DSP Code Generation", *Proc. of International Conference on Computer Aided Design*, November, 1996, pp. 109-112.

[61]   R. Leuper, P. Marwedel, "Retargetable Generation of Code Selectors from HDL Processor Models", *Proc of the European Design & Test Conference*, March, 1997, pp. 140-144.

148

[62] S. Liao et al., "Code Generation and Optimization Techniques for Embedded Digital Signal Processors", in Hardware/Software Co-design, ed. by M. Sami, G. DeMicheli, Kluwer Academic Publishers, 1996.

[63] S. Liao, S. Devadas, K. Keutzer, S. Tjiang, A. Wang, "Code Optimization Techniques for Embedded DSP Microprocesseurs", *Proc. of the Design Automation Conference*, 1995, pp. 599-604.

[64] C. Liem, M. Cornero, M. Santana, P. Paulin, A. Jerraya, J-M Gentit, J. Lopez, X. Figari, L. Bergher, "An Embedded System Case Study: the FirmWare Development Environment for a Multimedia Audio Processor", *Proc. of the Design Automation Conference*, Anaheim, CA, 1997.

[65] C. Liem, T. May, P. Paulin, "Instruction-Set Matching and Selection for DSP and ASIP Code Generation", *European Design & Test Conference*, Paris, France, Feb 1994, pp. 31-37.

[66] C. Liem, T. May, P. Paulin, "Register Assignment through Resource Classification for ASIP Microcode Generation", *Proc. of the Int. Conference on Computer Aided Design*, Santa Clara, CA, Nov. 1994, pp. 397-402.

[67] C. Liem, F. Naçabal, C. Valderrama, P. Paulin, A. Jerraya, "System-on-a-Chip Cosimulation and Compilation", *IEEE Design & Test of Computers, special issue on Design, Test, & ECAD in Europe*, June 1997.

[68] C. Liem, P. Paulin, M. Cornero, A. Jerraya, "Industrial Experience using Rule-driven Retargetable Code Generation for Multimedia Applications", *Proc. of the International Symposium on System Synthesis*, Cannes, France, Sept. 1995, pp. 60-65.

[69] C. Liem, P. Paulin, A. Jerraya, "Address Calculation for Retargetable Compilation and Exploration of Instruction-Set Architectures", *Proc. of the Design Automation Conference*, 1996, pp. 597-600..

[70] C. Liem, P. Paulin, A. Jerraya, "ReCode: the Design and Re-design of the Instruction Codes for Embedded Instruction-Set Processors", *Proc. of the European Design & Test Conference*, 1997.

[71] A. Lioy, M. Mezzalama, "Automatic Compaction of Microcode", *Microprocessors and Microsystems*, vol 14, no 1. January/February, 1990. pp 21-29.

[72] P. Lippens, J. van Meerbergen, A. van der Werf, W. Verhaegh, B. McSweeney, J. Huisken, O. McArdle, "PHIDEO: a silicon compiler for high speed algorithms", *Proc. of the European Design Automation Conference*, pp., 436-441, 1991.

[73] P. Marwedel, "Tree-based Mapping of Algorithms to Pre-defined Structures", *Proceedings of the International Conference on CAD*, 1993, pp. 586-593.

[74] P. Marwedel, "Tree-based Mapping of Algorithms to Predefined Structures (Extended Version)", *Report No. 431, University of Dortmund*, January, 1993.

[75] P. Marwedel, G. Goossens, editors, Code Generation for Embedded Processors, , Kluwer Academic Publishers, 1995.

[76] V. Milutinovic, editor, M. Flynn, foreword, High-Level Language Computer Architecture, Computer Science Press, 1989.

[77] E. Morel, C. Renvoise, "Global optimization by suppression of partial redundancies", *Communications of the ACM*, Vol. 22, No. 2, Feb. 1979, pp. 96-103.

[78] Motorola, "Motorola DSP Product Overviews", available at http://www.mot.com/SPS/DSP/home/prd/prodover.html

[79] R. Mueller, M. Duda, S. O'Haire, "A Survey of Resource Allocation Methods in Optimizing Microcode Compilers", *Proc. of the 17th Annual Workshop on Microarchitecture*, 1984, pp. 285-295.

[80] F. Naçabal, O. Deygas, P. Paulin, M. Harrand, "C-VHDL Co-Simulation: Industrial Requirements for Embedded Control Processors", *Proc. of EuroDAC/EuroVHDL Designer Sessions*, Geneva, Sept. 1996, pp. 55-60.

[81] S. Novack, A. Nicolau, N. Dutt, "A Unified Code Generation Approach using Mutation Scheduling", in Code Generation for Embedded Processors, ed. by P. Marwedel, G. Goossens, Kluwer Academic Publishers, 1995.

[82] P. Panda, N. Dutt, A. Nicolau, "Memory Organization for Improved Data Cache Performance in Embedded Processors", *Proc. of the International Symposium on System Synthesis,* 1996, pp. 90-95.

[83] P. Paulin, M. Cornero, C. Liem, F. Naçabal, C. Donawa, S. Sutarwala, T. May, C. Valderrama, "Trends in Embedded Systems Technology: An Industrial Perspective", in Hardware/Software Co-design, ed. by M. Sami, G. DeMicheli, Kluwer Academic Publishers, 1996.

[84] P. Paulin, J. Fréhel, M. Harrand, E. Berrebi, C. Liem, F. Naçabal, JC Herluison , "High-Level Synthesis and Codesign Methods: An Application to a Videophone Codec", *Proc. of EuroDAC/ EuroVHDL,* Sept. 1995.

[85] P. Paulin, J. Knight, "Force-Directed Scheduling in Automatic Data Path Synthesis", *Proc. of the Design Automation Conference,* 1987, pp. 195-202.

[86] P. Paulin, C.Liem, M.Cornero, F.Naçabal, G. Goossens, "Embedded Software in Real-time Signal Processing Systems: Application and Architecture Trends", *Proceedings of the IEEE, special issue on Hardware/Software Co-design,* 1997, pp. 444-451.

[87] P. Paulin, C. Liem, T. May, S. Sutarwala, "FlexWare: A Flexible FirmWare Development Environment", in Code Generation for Embedded Processors, ed. by P. Marwedel, G. Goossens, Kluwer Academic Publishers, 1995.

[88] N. Ramsey, D. Hanson, "A Retargetable Debugger", ACM *SIGPLAN Conference on Programming Language Design and Implementation,* Vol. 27, No. 7, July 1992, pp. 22-31.

[89] R. Rauscher, M. Koegst, "A System for Microcode Reduction", *Proc of the IFIP Int. Workshop on Logic and Architecture Synthesis,* Grenoble, France, 1996, pp. 379-386.

[90] K. Rimey, P. Hilfinger, "A Compiler for Application-Specific Signal Processors", VLSI Signal Processing III, IEEE Press, November 1988, pp. 341-351.

[91] SGS-Thomson Microelectronics, "D950-CORE Specification", January 1995.

[92] SGS-Thomson Microelectronics, "STi1100 Video CODEC Specification", August, 1993.

[93] SGS-Thomson Microelectronics, "STi3400 MPEG/H.261 Video Decoder Specification", January 1996.

[94] SGS-Thomson Microelectronics, "STi4600 6 channel Ddolby AC-3 MPEG 1/2 Audio Decoder Advance Data", January 1997.

[95] SGS-Thomson Microelectronics, "ST9 Family 8.16 bit MCU: Databook", July 1991, and "Technical Manual", November 1991.

[96] R. Stallman, "Using and porting GNU CC", Free Software Foundation, June 1994.

[97] M. Strik et al. , "Efficient Code Generation for In-House DSP-Cores", *Proc. of the European Design & Test Conference,* 1995, pp. 244-249.

[98] A. Sudarsanam, S. Malik, "Memory Bank and Register Allocation in Software Synthesis for ASIPs", *Proc. of the International Conference on CAD,* 1995, pp. 388-392.

[99] S. Sutarwala, P. Paulin, "Flexible Modeling Environment for Embedded Systems Design", *Proc. of the Int. Workshop on Hardware/Software Co-design,* Grenoble, France, 1994, pp. 124-130.

[100] Texas Instruments, "Digital Signal Processing Solutions", available at http://www.ti.com/sc/docs/ dsps/products.html

[101] H. Tomiyama, H. Yasuura, "Optimal Code Placement of Embedded Software for Instruction Caches", *Proc. of the European Design & Test Conference,* March, 1996, pp. 96-101.

[102] Understanding and Using COFF, O'Reilley & Associates, 196 p, see http://www.ora.com

[103] C.A. Valderrama, A. Changuel, P.V. Raghaven, M. Abid, T. Ben Ismail, A.A. Jerraya, "A Unified Model for Cosimulation and Cosynthesis of Mixed Hardware/Software Systems", *Proc. of the European Design &Test Conference,* March 1995, pp. 180-184.

[104] C.A. Valderrama, F. Naçabal, P. Paulin, A. Jerraya, "Automatic Generation of Interfaces for Distributed C-VHDL Cosimulation of Embedded Systems: an Industrial Experience", *Proc. of the International Workshop on Rapid Systems Prototyping*, June 1996, pp. 72-77.

[105] P. Vanoostende et al., "Retargetable Code Generation: Key Issues for Successful Introduction", in Code Generation for Embedded Processors, ed. by P. Marwedel, G. Goossens, Kluwer Academic Publishers, 1995.

[106] J. VanPraet, G. Goossens, D. Lanneer, and H. DeMan, "Instruction Set Definition and Instruction Selection for ASIPs", *Proc. of the International Symposium on High-Level Synthesis*, May 1994, pp. 11-16.

[107] S. Vercauteren, B. Lin, H. DeMan, "Constructing Application-Specific Heterogeneous Embedded Architectures from Custom HW/SW Applications", *Proc. of the Design Automation Conference*, June 1996, pp. 521-526.

[108] B. Wess, "On the Optimzal Code Generation for Signal Flow Graph Computation", *Proc. of the International Symposium on Circuits and Systems*, 1990, pp. 444-447.

[109] B. Wess, "Code Generation based on Trellis Diagrams", in Code Generation for Embedded Processors, ed. by P. Marwedel, G. Goossens, Kluwer Academic Publishers, 1995.

[110] T. Wilson, G. Grewal, S. Henshall, D. Banerji, "An ILP-based Approach to Code Generation", in Code Generation for Embedded Processors, ed. by P. Marwedel, G. Goossens, Kluwer Academic Publishers, 1995.

[111] R. Woudsma. et.al.. "EPICS, a Flexible Approach to Embedded DSP Cores", *Proc. of the International Conference on Signal Processing and Applications and Technology*, Dallas, TX, Oct. 1994.

[112] V. Zivojnovic et al, "DSPstone: A DSP-Oriented Benchmarking Methodology", *Proc. of the International Conference on Signal Processing and Technology*, Dallas, TX, Oct. 1994.

[113] V. Zivojnovic, J. Velarde, C. Schlager, "DSPStone: A DSP-Oriented Benchmarking Methodology", *Internal Report*, Aachen University of Technology, Aug. 1994.

[114] V. Zivojnovic, H. Meyr, "Compiled HW/SW Co-simulation", *Proc. of the Design Automation Conference*, Las Vegas, NV, June, 1996, pp. 690-695.

[115] Zoran Corporation, "ZR38500 Six-channel Dolby Digital Surround Processor: Preliminary Specification", November 1994.

[116] see ftp://ftp.prep.ai.mit.edu

[117] see http://www.mot.com

[118] see http://www.metaware.com

[119] see http://www.plumhall.com

[120] see htttp://www.research.ibm.com/vliw

[121] see http://www.ti.com

[122] see http://www.xemacs.org

[123] available from ftp://suif.stanford.edu/pub/tjiang.

[124] available from http://www.cs.princeton.edu/software/iburg

Glossary of Abbreviations

ACU = Address Calculation Unit
ALU = Arithmetic and Logic Unit
ASIC = Application Specific Integrated Circuit
ASIP = Application Specific Instruction-Set Processor
BDS = BNR Data Structure (an internal form for the CodeSyn compiler
developed at BNR/Nortel)
BNR = Bell-Northern Research (now Nortel)
BSP = Bit Stream Processor (an operator of the SGS-Thomson
Integrated Video Telephone)
CDFG = Control-Data Flow Graph
CISC = Complex Instruction-Set Computer
CODEC = Coder / Decoder
COFF = Common Object File Format (a debug format)
dag = directed acyclic graph
DAP = Digital Audio Processor (a processor developed at SGS-Thom-
son Microelectronics)
DCC = Digital Compact Cassette
DCU = Data Calculation Unit
DECT = Digital European Cordless Telephone
DMA = Direct Memory Access
DSP = Digital Signal Processor *or* Digital Signal Processing
DWARF = Debug With Arbitrary Record Format (a debug format)
ELF = Embedded Linker Format (a debug format)
FFT = Fast Fourier Transform
FIR = Finite Impulse Response (a type of DSP filter)
gcc = GNU C compiler
GNU = GNU's Not Unix (recursive definition)
GSM = Groupe Special Mobile (European cellular standard)
HDL = Hardware Description Language
ICE = In-Circuit Emulator *or* In-Circuit Emulation
IIR = Infinite Impulse Response (a type of DSP filter)
ILP = Instruction Level Parallelism *or* Integer Linear Program
IMEC = Interuniversitair Micro-Elecktronica Centrum /
Interuniversity Microelectronics Centrum
(a Belgian research institute)
INPG = Institue National Polytechnique de Grenoble /
National Polytechnical Institute of Grenoble

ISG	= Instruction Set Graph (a model used by the Chess compiler developed at IMEC)
ISO	= International Standards Organization
ITU	= International Telecommunications Union
IVT	= Integrated Video Telephone
MAC	= Multiply Accumulator
MAD	= Multiply Adder
MCU	= Microcontroller Unit
MI	= Micro-Instruction
MIPs	= Million Instructions Per Second
MIT	= Massachussettes Institute of Technology
MMDSP	= Multimedia Digital Signal Processor (a processor developed at Thomson Consumer Electronics Components)
MMIO	= Memory-Mapped Input / Output
MOP	= Micro-Operation
MPEG	= Motion Picture Experts Group
MSQ	= MicroSeQuencer (an operator of the SGS-Thomson Integrated Video Telephone)
nML	= not a Machine Language (an instruction-set specification language)
RAM	= Random Access Memory
RISC	= Reduced Instruction Set Computer
ROM	= Read Only Memory
RTL	= Register Transfer Level (a hardware description level) *or* Register Transfer Language (gcc internal representation)
SPAM	= Synopsys Princeton Aachen MIT (a joint compiler project)
ST	= SGS-Thomson Microelectronics
TCEC	= Thomson Consumer Electronics Components
TI	= Texas Instruments
TIMA	= Techniques de l'Informatique et de la Microélectronique pour l'Architecture d'ordinateurs / Techniques of Informatics and Microelectronics for computer Architecture (a French research laboratory)
VHDL	= VHSIC (Very High Speed Integrated Circuit) Hardware Description Language
VIP	= VLIW Image Processor (an operator of the SGS-Thomson Integrated Video Telephone)
VLIW	= Very Long Instruction Word
VLSI	= Very Large Scale Integration *or* Integrated Circuit
VOP	= Vertical Operation

Index